GENETICS - RESEARCH AND ISSUES

SEXUAL SELECTION

EVOLUTIONARY PERSPECTIVES, MATING STRATEGIES AND LONG-TERM EFFECTS ON GENETIC VARIATION

GENETICS - RESEARCH AND ISSUES

Additional books in this series can be found on Nova's website
under the Series tab.

Additional e-books in this series can be found on Nova's website
under the e-book tab.

GENETICS - RESEARCH AND ISSUES

SEXUAL SELECTION

EVOLUTIONARY PERSPECTIVES, MATING STRATEGIES AND LONG-TERM EFFECTS ON GENETIC VARIATION

ROBERT M. GELDANI
AND
MATTHEW A. DAVIN
EDITORS

NOVINKA

New York

LIBRARY OF CONGRESS CATALOGING-IN-PUBLICATION DATA

ISBN: 978-1-62808-805-2
Library of Congress Control Number: 2013946510

Published by Nova Science Publishers, Inc. † New York

Contents

PREFACE

In this book, the authors present current research in the study of the evolutionary perspectives, mating strategies and long-term effects on genetic variation of sexual selection. Topics discussed in this compilation include sexual dimorphism in insect longevity; evidence of natural and sexual selection shaping the size of nuptial gifts among a single bush-cricket genus; mate choice copying in both sexes of the guppy; and sexual selection under parental choice.

Chapter 1 – In species with separate sexes, gender differences in longevity are widespread and the extent and direction of these differences varies tremendously among taxa. To understand sexual dimorphism in longevity and explain how different forms of selection shape longevity and other fitness-related traits within and among species, it is important to obtain information on the genetic architecture (the number of genes and degree of inter- and intra-genic interactions) and various mechanistic causes which underlie mortality variation between the sexes. Here the authors review recent empirical studies on gender differences in longevity in insect species, from both mechanistic and evolutionary perspective. Whenever it was possible, they focus on data obtained from the laboratory evolution experiments because the study of evolution under controlled conditions may provide valuable evidence not only for the effects of natural and sexual selection in shaping sex-specific longevities and mortality rates but also it offers novel insights into the mechanistic basis of these differences.

Chapter 2 – During mating, male bush-crickets transfer a complex spermatophore to the female. The spermatophore is comprised of a large nuptial gift which the female consumes while the sperm from the ejaculate-containing ampulla are transferred into her. Two main functions of the nuptial gift have been proposed: the ejaculate protection hypothesis and the parental investment hypothesis. The former, founded on sexual selection theory, predicts that the time to consume the gift is no longer than necessary to allow for full ejaculate transfer. The latter maintains that gift nutrients increase the fitness or quantity of offspring and hence the gift is likely to be larger than is necessary for complete sperm transfer. With an aim to better understanding the primary function of nuptial gifts, the authors examined sperm transfer data from field populations of five *Poecilimon* bush-cricket taxa with varying spermatophore sizes. In the species with the largest spermatophore, the gift was four times larger than necessary to allow for complete sperm transfer and is thus likely to function as paternal investment. Species with medium and small gifts were respectively sufficient and insufficient to allow complete sperm transfer and are likely to represent, to various degrees, ejaculate protection. The authors also found that species that produce larger spermatophores transfer greater proportions of available sperm than species producing smaller spermatophores, and thus achieve higher paternal assurance.

Chapter 3 – Mate choice copying was mostly described as a strategy employed by females to assess the quality of potential mates, but also males can copy other males' mate choice. In both cases, focal individuals show an increased propensity to copulate with a potential mating partner they could observe interact sexually with another individual (the 'model'). Sexual interactions, however, convey additional—partly conflicting—information to the choosing individual: females may try to avoid sexually active males due to a rougher courtship or coercive mating attempts, and males may respond to an increased sperm competition risk (SCR). How do females and males respond to different copying situations, in which the model individual either resides in the vicinity of a potential mating partner without physical contact (i.e., no harassment, and low SCR), or interacts sexually with the potential mating partner? Do individuals copy less in the latter situation? The authors investigated these questions in the guppy (*Poecilia reticulata*), a livebearing fish with internal fertilization, strong SCR among males, and frequent sexual harassment of females. Focal individuals could choose to associate with a large or a small stimulus fish, and mate choice tests were repeated after the previously non-preferred stimulus fish could be seen associating (low harassment and low SCR) or physically interacting (high harassment and high SCR) with a model individual. In a control treatment, no model was presented. They found both males and females to copy similarly in both copying treatments, while no response was observed in the control. This contrasts with a study reporting that Atlantic molly (*P. mexicana*) males copy less under elevated SCR. Even though the authors lack a compelling explanation as to why both congeners might differ, they are tempted to argue that strong(er) benefits arising from copying may have selected guppies to copy in a broader range of contexts, including situations where the choosing individual incurs harassment or SCR.

Chapter 4 – Across pre-industrial human societies mating is regulated, with arranged marriage, in which male parents select spouses for their female relatives, being the primary mode of long-term mating. This chapter reviews the model of sexual selection under parental choice that offers a good account for these patterns of human mating. This model postulates that parent-offspring conflict over mating induces parents to control the mate choices of their children, and the spouses they select for them are individuals who conform best to their preferences. By doing so, parents become a significant evolutionary force affecting the course of sexual selection. The model is applied in order to achieve a comprehension of the evolution of specific mating strategies. In particular, it is argued that in a context where mate choice is regulated, at least three mating strategies can thrive: addressing parental choice, addressing female choice and circumventing parental and female choice by force.

In: Sexual Selection
Editors: R. Geldani and M. Davin
ISBN: 978-1-62808-805-2

Chapter 1

SEXUAL DIMORPHISM IN INSECT LONGEVITY: INSIGHTS FROM EXPERIMENTAL EVOLUTION

Jelica Lazarević*[*1], Biljana Stojković[2] and Nikola Tucić[1]
[1]Institute for Biological Research, University of Belgrade, Serbia
[2]Faculty of Biology, University of Belgrade, Serbia

ABSTRACT

In species with separate sexes, gender differences in longevity are widespread and the extent and direction of these differences varies tremendously among taxa. To understand sexual dimorphism in longevity and explain how different forms of selection shape longevity and other fitness-related traits within and among species, it is important to obtain information on the genetic architecture (the number of genes and degree of inter- and intra-genic interactions) and various mechanistic causes which underlie mortality variation between the sexes. Here we review recent empirical studies on gender differences in longevity in insect species, from both mechanistic and evolutionary perspective. Whenever it was possible, we focus on data obtained from the laboratory evolution experiments because the study of evolution under controlled conditions may provide valuable evidence not only for the effects of natural and

* To whom correspondence should be addressed: Jelica Lazarević, Institute for Biological Research, University of Belgrade, Despot Stefan Blvd. 142, 11000 Belgrade, Serbia, e-mail: jellaz@ibiss.bg.ac.rs.

sexual selection in shaping sex-specific longevities and mortality rates but also it offers novel insights into the mechanistic basis of these differences.

INTRODUCTION

Over the last 30 years our understanding of the proximate (mechanistic) causes involved in age-specific decline in survival and reproductive performance (senescence or ageing) has been improved dramatically. However, many fundamental questions concerning the relative importance of evolutionary mechanisms (i.e. ultimate causes: Mayr 1961) that directly or indirectly influence the rate of senescence are not entirely answered. For example, despite the fact that we now have a plethora of information with respect to longevity differences in genetic architecture and physiology between females and males (see below), ultimate causes why females and males age at different rates are still poorly understood.

Gender differences in longevity are widespread in insect species although the extent and direction of the difference depend on species phylogenetic position, environmental conditions and mating status (Table 1). Moreover, even when females and males have the same longevities, they may differ in demographic parameters, i.e., the shape of age-specific survival curves.

Antagonistic relationship between somatic maintenance and reproduction has been revealed by many longevity-extending genetic and environmental manipulations, as well as in artificial selection and laboratory evolution experiments (Hughes and Raynolds 2005). The cost of reproduction in many insects is frequently manifested as increased mortality rate of mated relative to unmated females. Mating is energetically costly to females not only because of high amount of resources allocated to egg production but also due to impairment of immune functions (Rolff and Siva-Jothy 2002; Fedorka and Zuk 2007), oxidative damage (Archer et al. 2012a) and direct physical harassment by males (Morrow and Arnquist 2003). In general, mated individuals live shorter than virgins, although many exceptions from this 'rule' (Lazarević 1994; Carey et al. 2002; De Loof 2011) imply that mating costs and benefits for female longevity may vary among different populations (Shuker et al. 2006) and species (Rönn et al. 2006). For example, multiple mating is beneficial for *Callosobruchus maculatus* (Rönn et al. 2006) and *Gryllus*

Table 1. Sexual dimorphism in insect longevity and demographic parameters depending on mating status

Species[1]	Mating[2]	Food[3]	Line, population	Females *vs.* Males					Reference
				Best fit model[4]	a	b	s (and/ or M)	Mean longevity	
Gryllodes sigillatus (O)	SM-I	+	8 inbred lines	G	<	=		<	Archer et al. 2012b
Teleogryllus commodus (O)	V	+	Up (selection for↑ male longevity)	G	=	=		>	Hunt et al. 2006
	V	+	Down (↓ male longevity)	G	=	<		>	
	MM	+	Seminatural cond.	GM	<	>	>	>	Zajitschek et al. 2009a
	MM♀, V♂	+		G	=	>		<	Zajitschek et al. 2009b
	MM♀, V♂	+/-		G	<	>		<	
Ceratitis capitata (D)	V-Gr	+		L	>	<	=	>	Pujol-Lereis et al. 2012
Drosophila melanogaster (D)	V-Gr	+	*Oregon* R	G	<	=		<	Vaiserman et al. 2013
	V-Gr	+	*Canton* S	G	<	=		=	
	V-Gr	+	*Um*	G	<	>		<	
	MM	+	FTF x Crete (Interpopulation crosses)	LM	<	>	> (<)	<	Rand et al. 2006
	MM	+	Crete x FTF (Interpopulation crosses)	LM	=	=	< (=)	<	
	V-I	+	*Canton* S	G	<	=		=	Iliadi et al. 2009
	SM-I	+		G	>	=		<	

Table 1. (Continued)

Species[1]	Matin g[2]	Food[3]	Line, population	Females *vs.* Males					
				Best fit model[4]	a	b	s (and/ or M)	Mean longevity	
	V-Gr	+		G	=	=		=	
	SM-Gr	+		G	=	=		<	
	MM	+		G	=	=		=	
	V-I	+	*Oregon* R	G	<	=		>	
	SM-I	+		G	=	=		=	
	V-Gr	+		G	<	=		>	
	SM-Gr	+		G	=	=		=	
	MM	+		G	=	=		=	
	V-Gr	+	Wt, peach orchard	GM	=	=	=	=	Promislow and Haselkorn 2002
Drosophila affinis	V-Gr	+	Wt, peach orchard	GM	<	=	>	<	
Drosophila hydei	V-Gr	+	*Drosophila* stock centre	GM	<	>	>	<	
Drosophila simulans	V-Gr	+	Wt, peach orchard	GM	<	>	>	=	
Drosophila virilis	V-Gr	+	*Drosophila* stock centre	GM	<	<	=	>	
Drosophila koepferae	V-Gr	+	Lab pop. (7gen.)	G	=	<		>	Scannapieco et al. 2007
Drosophila buzzati	V-Gr	+	Lab pop. (7gen.)	G	>	<		>	
	MM	+	control	G	=(<)	>		<	Scannapieco et al. 2009
	MM	+	L (selection for↑ lifespan) 30gen.	G	>	=		<	
	MM	+	Lab pop. (7gen.)	G	>	<		<	
	MM	+	Low latitude (natural population)	G	>	<		<	
	MM	+	High-latitude (natural population)	G	>	<		<	

Species[1]	Mating[2]	Food[3]	Line, population	Females vs. Males					
				Best fit model[4]	a	b	s (and/ or M)	Mean longevity	
Drosophila mojavensis	V-Gr	Agria	mainland Mexico (natural population)	G♀, GM♂	>	<		=	Jaureguy and Etges 2007
	V-Gr	Lab food	,,	L♀, LM♂	>	<	<	>	
Acanthoscelides obtectus (C)	V-I	-	E (early reproduction)	G	<	>		<	Stojković and Savković 2011
	V-I	-	L (late reproduction)	G	=	=		>	
	SM-I	-	E (early reproduction)	G	>	<		=	
	SM-I	-	L (late reproduction)	G	>	<		>	
	V-Gr		E (early reproduction)	G	=	<		>	Lazarević et al.. 2013
	V-Gr		L (late reproduction)	G	<	<		>	
Callosobruchus maculatus (C)	V-I	-	Vigna radiata pop	L	>	<	<	>	Fox et al. 2003
Stator limbatus (C)	V-I	-		L	<	=	=	=	
Pararge aegeria (Lep)	V-Gr	+	Madeira pop.	L	<	>	=	=	Gotthard et al. 2000
	V-Gr	+	Sweden pop.	L	>	<	<	>	

[1]O-Orthpoptera, D-Diptera, C-Coleoptera, Lep-Lepidoptera.

[2]Mating: I-individually kept; Gr-kept in group V-virgin; SM- single mated, MM- multiple mating.

[3]Adult food: + (*ad libitum*), - (without food), +/- dietary restriciton.

[4] L-logistic, G – Gompertz, GM - Gompertz-Makeham, LM – Logistic-Makeham.

lineaticeps (Wagner et al. 2001), detrimental for *C. chinensis* (Rönn et al. 2006), *Drosophila buzzati* (Scannapieco et al. 2007, 2009), *Acanthoscelides obtectus* (Šešlija et al. 2008), *Gryllus bimaculatus* (Bateman et al 2006), and does not affect female longevity in *Telostylinus angusticollis* (Adler et al. 2013) and *Gryllodes sigillatus* (Ivy and Sakaluk 2005).

Importance of genetic background in shaping insect mortality rates and sexual dimorphism in longevity was confirmed in many studies in which life-span of unmated (virgin) individuals were compared among different natural populations or laboratory lines (Yilmaz et al. 2008; Iliadi et al. 2009) and species of *Drosophila* (Promislow and Haselkorn 2002). *D. hydei* females live about 20% shorter than males, while life-span of *D. virilis* females is almost twice as long as male's (Promislow and Haselkorn 2002). *Canton S* and *Oregon R* lines of *D. melanogaster* differed in mean longevity of females, but not males, when they were kept individually, while in a single sex group of 20 individuals *Oregon R* females live longer and males live shorter comparing to *Canton S* line (Iliadi et al. 2009). Same authors showed that sexual dimorphism for longevity of virgin individuals was expressed only in *Oregon R* line, where females outlived males due to lower initial mortality rate. However, the results with *Oregon R* and *Canton S* lines may vary depending on experimental design (Cui et al. 2001; Moskalev et al. 2009; Vaiserman et al. 2013). As expected, QTL and quantitative genetic analysis revealed that genes affecting longevity are sex-specific and environment specific (Nuzhdin et al. 1997; Šešlija and Tucić 2008; Hallsson and Björklund 2011) implying that evolution of longevity may proceed by different mechanisms in females and males (Tower and Arbeitmann 2009) and may take different courses in diverse environmental circumstances.

Generally, life-span duration is influenced by female fitness-related traits more strongly compared with males, which is an expected consequence of positive correlation between female longevity and realized fecundity. Interestingly, this association between fecundity and life-span is even more important under stressful conditions. There is a plethora of evidence that females and males differently respond to the variations in environments. Substantial sex-specific plasticity (i.e., significant „sex × environment" interaction) in adult longevity has been detected in environments with varying nutritional value of diet (Adler et al. 2013; Zajitschek et al. 2009b), temperature (Norry and Loeschcke 2002), oxidative stress (Lazarević et al. 2013), etc. (reviewed in Section 3.3. of this chapter). As will be elaborated in proceeding sections of this chapter, different evolutionary mechanisms can be involved in shaping sexual dimorphism in longevity and patterns of ageing, as well as in disparity of environmental sensitivity of these characteristics among genders.

First, natural selection acting either directly on life-span duration in the two sexes or indirectly through environmental effects on other fitness-related traits (e.g. developmental time, body size, fecundity, etc.) can mould longevity patterns in specific habitats (Section 1). For example, in a lepidopterous species *Pararge aegeria*, temperate (Atlantic island of Madeira) and north populations (southern Sweden) exhibit phenological differences in adult eclosion. Since eclosion is more synchronized in north than temperate populations, northern females are available to males during a shorter period of time which weakens selective pressure for long life in males and promote higher sexual dimorphism in longevity (Gotthard et al. 2000).

Another evolutionary force which shape insect longevity and mortality rate is sexual selection (Section 2). Sexual selection occurs whenever sexes differ in investment to reproduction. Theory predicts that males adopt "live fast – die young" strategy while females follow "low-risk, low-wear-and-tear" strategy. Male-biased mortality is usually related to risky behaviour, such as male sexual competition in polygamous mating systems, and higher allocation of resources to reproduction and secondary sexual traits which improve sexual presentation. In monogamous species, or species where mating success increases with age, selection may favour lower mortality in males than females (Bonduriansky et al. 2008).

Effects of mating on longevity depend on mating system of a species. Comparisons of virgin and mated individuals between polyandrous and monoandrous species of fireflies from Lampyridae family revealed that in polyandry longevity was not affected by copulation while in monoandry single mated females doubled their lifespan relative to virgin counterparts (Fu et al. 2012). Analyses of fireflies from genus *Photinus* showed that, in comparison with males from monoandrous species, polyandrous males allocate more resources towards reproductive tissues, which maximize their ability to produce nuptial gifts and increase reproductive success under high levels of male-male competition (Demary and Lewis 2007). Among lepidopterous species, investment in early reproduction in females is negatively correlated with longevity and degree of polyandry possibly because polyandrous females obtain more nutrients through nuptial gifts than monoandrous ones (Jervis et al. 2005). Intra-species comparisons in *Acanthoscelides obtectus* pointed that monoandrous conditions, inadvertently created during laboratory evolution for early reproduction and short-life, decrease the magnitude of sexual dimorphism in longevity (Stojković and Savković 2011).

As stressed by Bonduriansky et al. (2008), females and males do not differ in the level but rather in the nature and scheduling of reproductive effort. In natural conditions, *Teleogryllus commodus* females live longer, show lower baseline mortality and higher mortality rate than males. At the same time, long life is related to early reproductive effort in females and late reproductive effort in males (Zajitschek et al. 2009a). Under laboratory conditions, similar age-dependent pattern of reproduction has been recorded, but males outlived females due to lack of extrinsic mortality causes (Maklakov et al. 2009b). This implies that environmental conditions determine optimal relationship between longevity and reproduction.

1. The Role of Natural Selection in Shaping Insect Longevity

Fundamental to the standard theory of ageing is the idea that senescence is a result of decreasing intensity of natural selection after the onset of reproduction (Medawar, 1952; Williams 1957; Hamilton 1966). This means that any delay in the start of reproduction should lead to a postponement of senescence. This prediction has received quite extensive support, primarily in the experimental evolution studies that have been conducted on three insect species, *Drosophila melanogaster* (e.g. Rose and Charelsworth 1980; Rose 1984; Partridge and Flower 1992), *Acanthoscelides obtectus* (Tucić et al. 1996, 2004) and *Musca domestica* (Reed and Bryant 2000). By manipulating the ages at which selection is strong (i.e. using early *vs.* late reproduction selection regimes), these studies have shown that natural selection is indeed able to shape the evolution of ageing patterns. Experimental evolution studies have also suggested that an important genetic mechanism underlying postponed senescence could be antagonistic pleiotropy, which assumes the existence of alleles that improve early-life performance but have deleterious effects at later ages. Because the intensity of natural selection declines with advancing age, these alleles will increase in frequency and cause the senescence due to their negative effects on late-age performance (Williams 1957). The observations of a trade-off between early and late life performance, as in the above mentioned studies, are empirical support of the antagonistic pleiotropy hypothesis. A second prediction of this hypothesis is the existence of negative genetic correlation between early and late-life fitness-related traits in populations with frequent segregating variation. These expectations also received experimental confirmation (e.g., Tucić et al. 1988), although in a less consistent way (Rose et al. 2005). It is noteworthy that an "early-late" trade-off in performance is also integral part of the "disposable soma" hypothesis (Kirkwood 1977), which proposes that senescence is caused by accumulation of damage due to processes that result in progressive decrease of the amount of resources allocated toward somatic maintenance and repair. This hypothesis assumes a physiological trade-off between reproduction and somatic maintenance and predicts that investment in reproduction (or somatic maintenance at earlier ages) should results in decrease of performance at later ages.

However, senescence might arise because of the accumulation of mutation with deleterious effects confined to later ages (Medawar 1952). Mutation accumulation hypothesis assumes that, because the strength of natural selection declines with age, genetic drift and recurrent deleterious mutations produce decline in individual performance at advancing ages. One of the effects of these processes could be upsurge of inbreeding depression for fitness-related traits with age. This is empirically testable prediction because crosses between populations, if they are subjected to age-specific mutation accumulation, should produce higher hybrid vigour (as the consequences of increased inbreeding load within populations) on fitness-related traits at later ages relative to the early ages (Charlesworth and Hughes 1996). Several studies of *Drosophila* have documented age-specific hybrid vigour for longevity and other fitness-related traits (Charlesworth and Hughes 1996; Hughes et al. 2002; Gong et al. 2006), but few others have shown that this relationship between age and inbreeding load varies among populations or are sex-specific (Lesser et al. 2006; Reynolds et al. 2007; Vaiserman et al. 2013). Furthermore, studies on two seed beetle species (*Callosobruchus chinensis*; Tanaka, 1990; and *C. maculatus*, Fox et al. 2006; Fox and Stillwell 2009) have produced results that were inconsistent with mutation accumulation hypothesis.

Experimental evolution studies, where replicate populations are exposed to different age-specific selection regimes, are especially suitable tool for testing mutation accumulation hypothesis. Here, the critical test of the hypothesis is the exploration of hybrid vigour in late-life fitness component in crosses between replicate populations subject to early reproduction. In study on short-lived early-reproducing *D. melanogaster* experimental populations, Rose et al. (2002) did not find differences between hybrid and nonhybrid populations in late-mortality rates implying that mutation accumulation could not be considered as a relevant explanation of genetic mechanisms involved in late mortality.

Also, in subsequent work of Borash et al. (2007) on the same set of fruit flies populations no detectable heterotic effect was seen for female and male longevities and their hybrids. The only fitness-related trait whose age-specific changes were consistent with mutation accumulation hypothesis was male mating ability. Interestingly, in the same kind of experiment on the seed beetle *Acanthoscelides obtectus*, in contrast to females, mating ability of males exhibited inbreeding depression (Đorđević et al., in preparation), indicating that underlying genetic mechanisms for reproductive effort differed between sexes in populations selected for early reproduction and short life. It is noteworthy that, in these nonhybrid populations of the seed beetle, virgin males outlived virgin females, while in the long-lived nonhybrid populations females lived longer than males (sex-reversal with respect to longevity in the short-lived populations was also observed in Stojković and Savković 2011). It seems that in short lived populations evolutionary course reversed sex-specific mortaliy rates being that in the ancestral experimental population (i.e., base population), from which short-and long-lived populations have been derived, the longevities of females were substantially higher than in males (Šešlija et al. 2008).

Although this finding of sex-specific responses to inbreeding is quite unexpected, it is consistent with the result obtained by Bilde et al. (2009) in another seed beetle, *Callosobruchus maculatus*.

As pointed out by Bilde et al. (2009), one possible explanation as to why inbreeding, which increases homozygosity, affects males less than females presumes that the heterogametic sex (males in *A. obtectus* and *C. maculatus*) does not suffer from increased expression of X-linked recessive deleterious mutations. Another potential explanation for the sex-specific sensitivity to inbreeding is that males evolve mechanisms to counteract the negative effect of mitochondrial mutations on longevity, possibly by nuclear genes that influence expression of the mitochondrial genes (Yee et al. 2013). This hypothesis is based on the fact that mitochondrial genomes commonly function sub-optimally when expressed in males, as a consequences of maternal inheritance of mitochondrial genes and and their selection within females.

Therefore, it is possible that the mismach between mitochondral and nuclear genes, which affect longevity, may be more pronounced in the inbred females than males. Consistent with this idea is the finding that in *D. simulans* females homozygous for some mutations in one nuclear-encoded mitochondrial protein pay a greater survival cost than do males (Melvin and Ballard 2011). The role of mitochondria in ageing process and sex-specific mortality patterns will be more elaborated in Section 3.3.1.

2. Sexual Conflict and Evolution of Sexual Dimorphism in Longevity

The opposing sex-specific responses to inbreeding observed in the two seed beetle species indicate that an optimal longevity has evolved not only as a results of trade-offs between reproductive performance early and late in life, but also that this process could be influenced by changes in patterns of sexual selection. Bilde et al. (2009) and Maklakov and Lummaa (2013) argue that it is likely that sex-specific selection, which assumes that the optimal reproductive strategies of females and males differ, has also played an important role in the evolution of sexual dimorphism in longevity and mortality rates. Sexually divergent selection may produce two distinct forms of conflict with very different evolutionary outcomes. Intra-locus sexual conflict, which occurs whenever reproductive behaviours that are optimal for one sex are deleterious for the other sex, prevents traits which are shared by the sexes to evolve to their sex-specific phenotypic optima, because the shared traits have a common genetic basis. In contrast, inter-locus conflict involves different genes in each sex and it can cause open-ended cycles of adaptation and counter-adaptation between the sexes (Parker 1979). Since intra-locus conflict could lead a subset of genes coding for the shared traits to sex limited expression (see e.g. Lande 1980; Bonduriansky 2007 and references therein), it is expected that traits under this form of sexual selection possess the sex-specific genetic architecture. This prediction is supported in both seed beetle species, *C. maculatus* (Fox et al. 2004) and *A. obtectus* (Šešlija and Tucić 2008). In both studies it has been shown that genes coding for extended longevity were dominant in females and recessive in males, which raises the possibility that inbreeding could decrease female longevity and, at the same time, increase life span in males. Thus, it seems that the genetic architecture of longevity in these species evolves in a way that reduces genetic constraints on the independent evolution of the sexes. In one experiment on *D. melanogaster*, Wayne et al. (2007) found that variation in gene expression was mostly the result of non-additive interactions between genes in females, while in males additive interactions of alleles prevailed. Having in mind that additive genetic variation respond to selection more readily, it is possible that genes affecting longevity may be more exposed to selection in males than in females, thereby reducing inbreeding load for male-specific deleterious mutations in populations.

Comparison of laboratory strains of *D. melanogaster* revealed that large part of transcriptional variation can be attributed to sexual dimorphism especially for genes expressed in germ-line tissues (Jin et al. 2001; Wayne et al. 2007; Baker et al. 2007; Wilson et al. 2013). These studies, in which sex-specific differences in the mode of transcriptome inheritance were identified, provide an additional insight into the gender differences in genetic architecture of the fitness-related traits. Gene expression pattern changes in response to mating and sexual selection and depending on mating system can have different evolutionary outcomes in females and males (Mank et al. 2013). For example, in *Drosophila* males, higher changes in transcription level during ageing was revealed for genes affecting protein synthesis, whereas in females starch and sucrose metabolism was more affected (Wilson et al. 2013). Magwire et al. (2010) analyzed effects of longevity-extending mutations on transcriptional pattern and found that each mutation had at least one deleterious pleiotropic effect and that these effects differed between the sexes.

The key feature of intra-sexual conflict in the evolution of longevity and ageing is that selection is working in the opposite directions in females and males, i.e. when female fitness is associated with increased longevity, male longevities should co-evolve, because of the intersexual genetic correlation for longevity, resulting in decreased male fitness. The predicted sex-specific relationship between fitness and longevity has been obtained in the seed beetle *C. maculatus*, where divergent sex-specific selection on longevity has been conducted (Berg and Maklakov 2012). In this experimental evolution study, selection for long-life produced replicate populations with low male fitness and high female fitness. More results implying that different reproductive strategies may lead females and males towards different optima in longevity has also been shown in the decorated cricket, *Gyllodes sigillatus* (Archer et al., 2012b). The sexual dimorphism in this species arises because females and males alter their reproductive effort with age; the reproductive effort increases with age in males but not in females. These strategies have led to an interesting evolutionary outcome - males live longer and age more slowly than females. Consistent with the view that intra-locus sexual conflict has the potential to play an important role in the evolution of sex differences in longevity and ageing are also findings obtained by Lewis et al. (2011) on indian meal moth, *Plodia interpunctella*. By using quantitative genetic approach these authors have shown that the pattern of multivariate selection acting on longevity and two other sexually dimorphic fitness-related traits was working in the opposing directions in the two sexes.

Several recent studies have indicated that some sexual adaptations in males, such as genital armour or increased toxicity of seminal fluid that polygynous males evolved to maximize their sperm competitive ability (possibly by destroying or disabling the sperm of previous mates), should lead to evolution of accelerated senescence in females. For example, in the laboratory evolution experiment on *C. maculatus*, Maklakov et al. (2007) have shown decreased rates of mortality and elevated longevity in monogamous virgin females relative to virgin females evolved under enforced polyandry. Similarly, female *D. melanogaster* had longer life-span after mating with males from lines selected for monogamy relative to females mated with polyandrous males (Holland and Rice 1999). Thus, removal of sperm competition may have promoted evolution of more benign mates for females, which liberated them from costly counter-adaptations. Also, under such conditions, males did not trade-off resources between traits associated with male-to-male competition and longevity. Monogamous and polyandrous mating regimes have also been applied in other insect species: *D. pseudoobscura* (Bacigalupe et al. 2007), the dung flies, *Sepsis synipsea* and *Scathophaga stercoraria* (Hosken et al. 2001; Martin and Hosken 2003), the dung beetle, *Onthophagus taurus* (Simmons and Garcia-Gonzalez 2008). These studies support the idea that sexual selection can lead to significant gender differences in longevity and mortality rates.

Despite the potential importance of combined effect of sexual selection and natural selection for the evolution of longevity and mortality rates, there are few empirical studies that have attempted to tease apart these two evolutionary mechanisms. Maklakov and Fricke (2009), Maklakov et al. (2009a) and Maklakov et al. (2010) pioneered the approach of exploring consequences of simultaneous manipulations of mating systems (monogamy *vs.* polyandry) and age-specific selection (early *vs.* late reproduction) on the evolution of longevity in *C. maculatus*. The results showed that natural selection affects the evolution of longevity much more than did sexual selection.

The general effects of the conflict between members of the same sex over opportunity to mate on the evolution of longevity and mortality rates can mediate, to some extent, the species-specific adaptations. For example, because mated females with repressed oviposition live longer than virgin females, it seems that seed beetle (*A. obtectus*) females under starvation benefit from nutrients and/or water received with ejaculates (Tucić et al. 1996; Maklakov et al. 2005; Šešlija et al. 2008). Since *A. obtectus* is capital breeder (adults do not need water or food to reproduce successfully), this can be treated as an important female adaptation. Similar effects of male ejaculates on female fitness have also been observed in other bruchid insects (Savalli and Fox 1999; Edvardsson 2007).

To date, the role of sexual conflict in shaping longevity has been tested in very few species and focusing mostly on the consequences of conflict from the female's perspective. From the male's point of view, Hall et al. (2009) have revealed that nuptial feeding could be an important mechanism of antagonistic interactions between females and males and that in some cases, when resources are scarce, male reproductive investment in offspring, together with the associated costs of mating, can exceed the costs experienced be females. Under these circumstances, as pointed out by Hall et al. (2009), "the sex roles can reverse with females actually competing for access to males as potential food sources" (p. 873).

In other words, changes in nature of sexual conflict may shape longevity and mortally rates of males in the same way as elevated sexual conflict do in females.

3. Mechanistic Causes of Sex-Specific Ageing

Many experiments have shown that various genetic and environmental interventions exhibit sex-specific effects on longevity. Because of these gender differences Burger and Promislow (2004) emphasized the importance of studying mechanisms of aging separately in females and males (Table 2).

Proximate causes of sex-specific variation in longevity, which are the fuel for natural and sexual selection in shaping ageing patterns across sexes, include, besides genetic variability, hormonally regulated differences in behaviour, metabolic and immune functions, as well as tolerance to oxidative and other stresses between females and males (May 2007). In addition, it is important to keep in mind that differential allocation of resources between longevity and reproduction also depend on developmental stage and environmental conditions (Boggs 2009).

Table 2. Relative effect (% of change relative to control) of different gene manipulations in *Drosophila* females and males. With the exception of Cox7A where effects of gene mutation were examined for *D. simulans* all other data are for *D. melanogaster*

Gene	Function	Manipulation	Specificity	Trait	conditions	Females	Males	
Cox7A homozygotes	Cytochrome c oxidase subunit	Hypomorphic mutation	Ubiquitous in females, testis specific in males	50% survival,	virgin 23°C, 50% humidity	-18	No effect	Melvin and Ballard 2011
Methuselah homozygotes	Cellular signalling	Hypomorphic mutation		Mean longevity *vs.* w^{1118} control	virgin dark, 29°C,75-90% humidity	+18	+15	Mockett and Sohal 2006
				Mean longevity *vs.* w^{1118}control	virgin 25°C,50% humidity	+5	+22	
Hsf4 homozygotes	Heatshock response			Mean longevity *vs.* *Canton*-S control	Virgin 25°C	-37	-33.5	Moskalev et al. 2009
Hsf2/CyO						-9	+10	
Canton-S × Hsf2/CyO						+33	+12	
Hsf^0		Heat-sensitive		Mean longevity *vs.* rescued mutant	Virgin, 25°C	+28	No effect	Sørensen et al. 2007
					Virgin, 25°C, high lipid food	-26	-29	
UAS-D-GADD45 × GAL4-1407	Stress response, DNA repair, apoptosis	conditional overexpression	Nervous system, adults	Mean longevity *vs.* UAS-D-GADD45parent	Males and non-virgin females	+25.5	+73**	Plyusnina et al. 2011
				Mean longevity *vs.* GAL4-1407 parent		+58	+71	
p53, homozygotes	Apoptosis, mitochondrial metabolism	null mutant	-	Mean longevity	Virgin, 25°C	+13	+12	Waskar et al. 2009
p53		conditional overexpression	whole tissue	Mean longevity	29°C	-4 to -6	+10 to +18	Shen and Tower 2010
p53		conditional overexpression	Nervous system			+11, +12	-9, -13***	
GLaz -/-, Nlaz+/+	Stress response, fat storage	null mutant	Nervous system	50% survival *vs.* GLaz +/+, Nlaz+/+	Virgin, 25°C, 60% humidity	No effect	-15.5	Ruiz et al. 2011

Table 2. (Continued)

Gene	Function	Manipulation	Specificity	Trait	conditions	Females	Males	
GLaz +/+, Nlaz-/-	Stress response, systemic inhibition of IIS	null mutant	Nervous system	„		-30.8	-22.5	
Human SOD	Antioxidative defence	overexpression	motorneurons	Mean longevity *vs.* UGA UAS-HS control strains	Virgin, 24°C	+43	+28	Spencer et al. 2003
InRp5545/ InRE19	Insulin-like receptor	hypomorphic		Mean longevity	Mated, 25°C	+85	No effect	Tatar et al. 2001
Chico homozygous mutant	Substrate for InR	hypomorphic		Life expectancy	Mated, 25°C, 40% humidity	+57.6	+5.5	Tu et al. 2002
Chico heterozygous mutant						+36	+50	
dFOXO homozygous	Transcription factor in IIS/IGF-1 pathway	null mutant	-	Mean longevity *vs.* Canton S	Virgin, 25°C	-51	-68	Moskalev et al. 2011
Lnk^{Del29} homozygotes	Insulin/IGF-1 signaling	loss-of-function		50% survival *vs.* w^{1118} control	Mated, 25°C	+11.5	+13	Slack et al. 2010
Lnk^{Del29} heterozygotes						-3	No effect	
Sirt2 homozygous	protein deacetylase	null	-	Mean longevity *vs.* Canton S	Virgin, 25°C	+30	-27	Moskalev et al. 2011
$Sir2^{EP2300}$	protein deacetylase	overexpression	ubiquitous	50% survival	Virgin, 25°C	+57	+32	Rogina and Helfand 2004
		overexpression	Nervous system			+52	+20	
		overexpression	Fat body	50% survival	Mated, 25°C	+12.3	+13	Hoffmann et al. 2013
Enigma heterozygotes	β oxidation of fatty actds	null mutant		50% survival vs. w^{1118} control	25°C	+19.5	No effect	Mourikis et al. 2006
Gr63a homozygotes	Receptor in olfactory neurons	null mutant		Mean longevity *vs.* w^{1118} control	Virgin, 25°C, 60%humidity	+30	No effect	Poon et al. 2010

Gene	Function	Manipulation	Specificity	Trait	conditions	Females	Males	
Tudor(1) homozygotes	Sexual differentiation	null mutant		Mean longevity *vs.* *Oregon* R control	Germ-line ablated, 25°C	-21,-2	+20, +20***	Shen et al. 2009
Snazarus X linked,	Endocytosis, signal transduction	transgene	fat body	Mean longevity *vs.* w^{1118} control	Virgin, 22-23°C	+62	+85	Stenesen 2011
Sirt2 homozygous	protein deacetylase	null	-	Mean longevity *vs.* Canton S	Virgin, 25°C	+30	-27	Moskalev et al. 2011
Sir2 ^{EP2300}	protein deacetylase	overexpression	ubiquitous	50% survival	Virgin, 25°C	+57	+32	Rogina and Helfand 2004
		overexpression	Nervous system			+52	+20	
		overexpression	Fat body	50% survival	Mated, 25°C	+12.3	+13	Hoffmann et al. 2013
Enigma heterozygotes	β oxidation of fatty actds	null mutant		50% survival vs. w^{1118} control	25°C	+19.5	No effect	Mourikis et al. 2006
Gr63a homozygotes	Receptor in olfactory neurons	null mutant		Mean longevity *vs.* w^{1118} control	Virgin, 25°C, 60%humidity	+30	No effect	Poon et al. 2010
Tudor(1) homozygotes	Sexual differentiation	null mutant		Mean longevity *vs.* *Oregon* R control	Germ-line ablated, 25°C	-21,-2	+20, +20***	Shen et al. 2009
Snazarus X linked,	Endocytosis, signal transduction	transgene	fat body	Mean longevity *vs.* w^{1118} control	Virgin, 22-23°C	+62	+85	Stenesen 2011

**average effect of 3 replications, ***two independent transgenes.

3.1. Neuroendocrine Regulation of Ageing

Neuroendocrine system regulate interaction between nutrition, metabolism, reproduction and ageing and adjust organism behaviour and physiology according to developmental and ecological context. This system of regulation presumes complex network of large number of interdependent elements whose empirical investigation requires thorough biochemical analyses in each population/ species/environment arrangement. Among various signalling pathways and gene functions involved in ageing-pattern regulation, here we elaborate several mechanisms that were extensively explored in insects.

Insulin signalling pathways link longevity, oxidative stress resistance and immunity. In *Drosophila*, down-regulation of insulin/IGF-1 signalling in mutants or individuals with ablated medial neurosecretory neurons (MNSN) increases longevity due to changes in the level of secondary hormones - juvenile hormone and ecdysone (Tatar et al. 2003). These hormones affect many aspects of reproduction such as oogenesis and spermatogenesis, synthesis of vitellogenin and accessory gland proteins, courtship behaviour, expression of secondary sexual traits etc. (Gruntenko and Rauschenbach 2008; Ishimoto et al. 2009; Moczek 2009; Toivonen and Partridge 2009). As expected, mutations in insulin/IGF-1 signalling pathway affects females more strongly than males since their investment in reproduction is higher (Clancy et al. 2001; Tatar et al. 2001; Tu et al. 2002). Ablation of MNSN increased median lifespan by 10.5% in males, 18.5% in virgin and 33.5% in mated females (Broughton et al. 2005). Beside reduced fecundity, mated females with ablated MNSN, also show increased lipid and carbohydrate storage, increased resistance to starvation and oxidative stress and reduced tolerance to heat and cold. The observed effects of insuline/IGF-1 signalling pathway on ageing are associated with its influence on levels of expression of some genes. For example, down-regulation of this signalling pathway prevent phosphorylation of dFOXO transcription factor and enables its translocation to nucleus leading to increased transcription of genes such as those related to oxidative stress resistance and response to nutrient variation. On the other hand, overexpression of dFOXO increases lifespan (Hwangbo et al. 2004), while null mutants for *dFOXO* gene exhibit significantly shorter life-span in both females and males (Moskalev et al. 2011). The importance of dFOXO activity for determining ageing pattern has been described in many studies. This transcription factor is involved in hormetic response to some stresses (Jünger et al. 2003; Moskalev et al. 2011), and also, by its influence on transcription of other genes, such as suppression of *Lnk* gene, dFOXO is involved in regulation of insulin/IGF-1 signalling pathway (Slack et al. 2010). Homozygous mutants for *Lnk* gene have increased lifespan in females and males, but females are more resistant to starvation and oxidative stress, and accumulate more lipids and trechalose than males (Slack et al. 2010).

dFOXO transcription factor, together with p53, control the expression of stress signalling gene *D-GADD45*. Its ubiquitous overexpression is letal (Peretz et al. 2007). However, in nervous system it extends lifespan without reduction in fecundity and locomotor activity (Plyusnina et al. 2011). Moreover, fecundity was higher which might explain differences in the mutation effects on females and males (Table 2). Another example of gene expressed in nervous tissue, which differently affects female and male longevity, is *GLaz* gene involved in regulation of stress resistance and fat storage. Null mutants showed reduced longevity only in males and this sex-specific modulation of longevity is related to changes in protein homeostasis (Ruiz et al. 2011). Mutant flies also showed reduced fecundity, deficient courtship and lower mating success. On the other hand, mutation in the gene for olfactory receptor increases both lifespan and reproduction in females, as well as fat storage and oxidative stress resistance (Poon et al. 2010).

As mentioned previously, insuline-like pathways control juvenile hormone and ecdysone - two signalling molecules that were found to be significant for sex-specific development of ageing patterns. Juvenile hormone is a pro-ageing hormone known to suppress stress resistance and immunity (Flatt et al. 2005) and to mediate trade-off between reproduction and immune functions. Addition of methopren (juvenile hormone analog) in larval food increases early fecundity and decreases lifespan in *D. melanogaster* (Flatt and Kawecki 2007). Mutations in ecdysone receptor (EcR) are lethal in homozygotes while heterozygous mutants show increased longevity by 45% both in females and males (Simon et al. 2003). However, mutant females exhibit higher resistance to heat stress, while males become more resistant to oxidative stress and dry starvation. In one experiment where *Drosophila* populations were selected for postponed reproduction, it was found that young females have decreased levels of 20-OH-ecdysone, while in males activity of this hormone was unaffected by selection regime (Harshman 1999). Similarly, higher investment in reproduction in short winged cricket females was related to increase in 20-OH-ecdysone, whereas short- and long-winged males did not differ in the level of 20-OH-ecdysone (Zera et al. 2007).

Together with juvenile hormone and ecdysone, another product of neurosecretory neurons that is involved in regulation of reproduction and mating behaviour is dopamine neurotransmitter (Gruntenko and Rauschenbach 2008). In addition, it seems that dopamine system also has a great impact on shaping longevity. Namely, it has been found in natural *Drosophila* populations that polymorphism in Dopa decarboxylase, enzyme of the last step of dopamine synthesis, account for a large fraction of genetic variation in longevity (De Luca et al. 2003). As shown, the level of dopamine is negatively correlated with lifespan (Vermeulen et al. 2006a). However, in response to selection for short life-span increased dopamine level was recorded in females, while in males the same dopamine response was obtained in populations selected for long life-span.

3.2. Metabolic Differences between Sexes and Longevity Consequences

Reallocation of limited resources among different functions underlies longevity-reproduction trade-off. However, in many species this physiological trade-off may not be obvious or complete being that different functions may require qualitatively different compounds for their energy supply. For example, although short-winged morph of *Gryllus* allocates more resources to ovaries comparing to long-winged morph, there is negative correlation between tryglyceride and phospholipide accumulation since tryglicerides are main flight fuel while phospholipids are main component of oocytes (Zera 2005).

Furthermore, strategy of metabolite accumulation during preadult development and their usage during ageing may differ between sexes. In *Ceratitis capitata* females and males differ in the pattern of change in lipid content with age (Pujol-Lereis et al. 2012). In *D. melanogaster* lipid content increases with ageing while opposite trend was recorded in males (Vermeulen et al. 2006b). On the other hand, in species which do not feed at adult stages, all metabolites decrease during ageing (Lazarević et al. 2012).

QTL analysis on *D. melanogaster* documented that lipid storage is under control of sex-specific genetic mechanisms (De Luca et al. 2005). Comparisons among different *Drosophila* species demonstrated significant among-species differences in sexual dimorphism of metabolite pool content (Matzkin et al. 2009). As shown on *D. melanogaster*, males have, on average, more tryacylglycerols than females (De Luca et al. 2005). Selection for postponed reproduction in the same species increased carbohydrate and lipid content in females, while in males only lipid content was increased (Djawdan

1996). In accordance with metabolite pool changes, both sexes in long-lived flies showed increased starvation resistance. However, only long-lived females were more resistant to desiccation comparing to control population (Djawdan 1998). In *Acanthoscelides obtectus*, on the other hand, selection for postponed reproduction and long life resulted in increased carbohydrate and lipid content in males, whereas in females only glycogen content was elevated (Lazarević et al. 2012). At the same time, selection for early reproduction and short life-span resulted in females with lower lipid and higher protein contents. Similar trade-off pattern was found for short-winged early-reproduction female crickets. Results of biochemical analyses on this species suggest that these changes are accompanied with alterations in activity of enzymes of intermediary metabolism (Zera 2005).

3.2.1. Nutritional Basis of Ageing

Studies on the nutritional basis of ageing revolve around positive effects of dietary restriction on life-span (Partridge et al. 2005; Piper and Partridge 2007) and the role of specific macronutrients and their ratio in shaping patterns of ageing and reproduction (Lee et al. 2008; Maklakov et al. 2008; Maklakov et al. 2009b; Simpson and Raubenheimer 2009). Generally, most studies have documented that two sexes differ in their sensitivity to diet quality and quantity.

Due to their higher investment in reproduction, it is expected that females are more responsive to nutritionally poor food than males (Nakagawa et al. 2012). Dietary restriction (40% dilution of sucrose-yeast medium) in Dahomey wild type virgin females of *Drosophila melanogaster* led to 60% increase in longevity while males lived only 30% longer (Magwere et al. 2004). However, when mated individuals of *Canton* S *Drosophila* line were fed on corn meal-sucrose-yeast medium, no sex differences were recorded (Min and Tatar 2006). Nutrient poor diet in *Telleogrylus commodus* did not affect male longevity while in females longevity was increased for about 60% (Zajitschek et al. 2009b). On the other hand, in *Telostylinus angusticollis*, both sexes elevated life-span for 65% on nutrient poor diet (Adler et al. 2013). Since life-extending effects could be achieved in female *D.melanogaster* without ovaries (Mair et al. 2004) it was suggested that reallocation of resources to reproduction is not the main cause of observed effects.

As stressed by Piper et al. (2011), exploring the effects of dietary balance is more informative than dietary dilution. Insects encountering imbalanced food can rebalance nutrient intake and utilization by behavioural and physiological mechanisms (Bede et al. 2007; Behmer 2009; Clissold et al.

2010). Higher consumption of nutritionally poor food has been described as compensatory response in many insects. However, in experiment of Min and Tatar (2006) males did not show compensatory response and females even increased intake of nutrient rich food. As for most other physiological processes, regulatory capacity, nutrient requirements and age-specific changes in nutrient needs may differ between the sexes. For example, behavioural adjustment to sex-specific dietary requirements could be achieved by sex-specific dietary preference. Female and male larvae of the gypsy moth in advanced instars differ in preference for lipids and proteins (Stockhoff 1993). When they were offered two complementary cubes of artificial diet, females chose to eat more proteins and males consume more lipids. This finding is in line with high energy demands in flying adult males, on one hand, and high investment of proteins in egg production in females, on the other hand. In cockroaches, males prefer carbohydrate biased food which increases their weight gain and lipid accumulation. In addition, this specific diet increases attractiveness and pheromone production in males implying that levels and courses of sexual selection may be largely influenced by dietary manipulations (South et al. 2011). Similarly, in *Telleogrylus commodus*, females prefer food with higher protein to carbohydrate ratio than males, which mirrors sex-specific difference in nutrient requirements for reproduction (Maklakov et al. 2008). However, in contrast to males, preferred food does not maximize reproduction in females implying that there might be intra-locus sexual conflict over nutrient regulation. Moreover, life-span and reproductive effort are maximized at different protein to carbohydrate ratios in females and males. In addition to the observed sex differences, Maklakov et al. (2008) also found that nutrient requirements for maximizing longevity differ from optimal nutrition for maximizing reproduction. In males, nutrient requirements for the two fitness components lay in the similar region of two-dimensional nutrient space, but in females, reproduction is maximized in much higher protein to carbohydrate ratio than longevity indicating that nutrient intake should make compromise between these two traits. Discrepancy in nutrient requirements between longevity and reproduction in females has also been recorded in *Drosophila melanogaster*, *Bactrocera tryoni*, *Anastrepha ludens* (reviewed in Simpson and Raubenheimer 2009) and explain shorter female life-span compared with males in these species. In difference to virgin males in the study of Maklakov et al. (2008), mated *T. commodus* males have different nutrient landscapes for longevity and reproductive effort (Archer 2012). It seems that, in presence of females, males have higher demands for proteins because of increased calling efforts, increased sperm production and production of spermatophore. When given a choice, mated males eat even more proteins than females. Geometric framework approach was also applied for studying effects of nutrient ratio on immune functions and it was shown that different functions (e.g., phenoloxidase and lysozyme activity) are maximized at different ratios of proteins and carbohydrates (Cotter et al. 2011).

Among genes that were found important for dietary restriction response (Banerjee et al. 2012) and fat metabolism (Reis et al. 2010) in insects, detail analysis was performed on protein deacetylase, sirtuin Sir2. Its overexpression in nervous system increased life-span more in females than males (Rogina and Helfand 2004). Overexpression of Sir2 in fat body provokes similar increase in longevity in both sexes, although sex-specific changes in transcriptional pattern of some genes were recorded (Hoffmann et al. 2013). The role of Sir2 in regulating longevity can be understood in the light of its effect on p53 transcriptional factor which is downstream target of Sir 2 (Bauer et al. 2009).

3.3. Stress Resistance and Life-Span

There is a significant overlap in gene expression patterns between stress response and ageing which point to involvement of stress resistance mechanisms in somatic maintenance (Vermeulen and Loeschcke 2007). Besides, selection for extended longevity is often related to increased stress resistance (Wit et al. 2013).

Exposure to a stressor activates various protective responses in cells and organism which may act as anti-ageing agents (Demirović and Rattan 2013). This hypothesis is confirmed by various beneficial effects of mild stresses, i.e., "hormesis", that were found in many animal species (see Rattan 2012). Hormetic effects, as other explained mechanisms involved in determination of life-span, were also shown to be sex-specific and dependent on dose and genetic background (Sarup and Loeschcke 2011). Generally, hormesis has been documented more frequently in males than females (Minois 2000). Male longevity was increased for hypergravity and cold (Le Bourg et al. 2009), X-radiation (Vaiserman et al. 2003), and heat (Sørensen et al. 2007). On the other hand, female *D. melanogaster* needed higher doses of morphin chloride for life-extending effect (Dubiley et al. 2011).

In *Drosophila melanogaster*, females are more resistant to desiccation (Gibbs et al. 1997; Chippindale et al. 1998), starvation (Harshman et al. 1999; Le Bourg 2007a), oxidative stress (Le Bourg 2007b; Minois 2001; Moskalev et al. 2009), high temperature (Le Bourg 2007a), and cold stress (Norry and Loeschke 2002). On the other hand, they are more sensitive to the presence of galactose in food (Cui et al. 2004), hypergravity (Le Bourg et al. 2009), radiation (Vaiserman et al. 2003), fungus infection (Le Bourg 2009) and combined effects of the stresses (Le Bourg 2012). In *Drosophila buzzatti* selected for increased longevity, increased heat stress resistance was found only in females (Scannapieco et al. 2009).

Investigations on the mechanistic basis of ageing focus mostly on the relationship between longevity and oxidative stress resistance. Even under optimal conditions, metabolic functions are related to production of reactive oxygen species (ROS) in mitochondria which may damage macromolecules and accelerate ageing if they are not balanced with ROS scavenging and damage repair systems (Isaksson et al. 2011). Despite many experiments which have succeeded to confirm basic postulates of free radical / oxidative stress theory of ageing (e.g., Arking et al. 2000, 2002; Wang et al. 2004; Orr et al. 2005) there are also many of them which failed to find positive correlation between oxidative stress resistance and longevity (Bayne et al. 2005; Loder 2006), between ROS production and longevity (Miwa et al. 2004; Sanz et al. 2010) or between activity of antioxidative enzymes and oxidative stress resistance (Seto et al. 1990; Mockett et al. 2001). It appeared that antioxidative defence is more relevant for survival under stressful conditions than for natural variation in ageing and longevity (Le Bourg 2001; Haenold et al. 2005). It must be kept in mind that reactive oxygen species, besides their detrimental effects, also have crucial regulatory role in gene expression, cell signalling, differentiation and apoptosis (Sohal and Orr 2012). They are involved in immune defence and mediate trade-off between immune function and somatic growth as well as between immunity, sexual ornamentation and sperm quality (Dowling and Sommons 2009). Thus, it is not surprising that many studies have revealed duplicitous role of oxidative stress determined by dose dependence of positive or negative effects of many antioxidant and prooxidant chemicals (Bahadorani et al. 2008; Bakkali et al. 2008; Ernst et al. 2013; Lazarević et al. 2013).

Depending on stress intensity (i.e., dose) and species' life-history strategies, oxidative stress can mediate trade-off between longevity and reproduction either as a consequence of differential allocation of antioxidants between soma and reproductive tissues or due to higher energy demands of increased reproductive effort (Monaghan et al. 2009). It is well known that oocytes accumulate antioxidant proteins such as superoxide dismutase, catalse, glutathione S transferase, yolk proteins (Kurama et al. 1995; Collins et al.

2004; Freitas et al. 2007; Diaz-Albiter et al. 2011) which can be used for defence against oxidative challenge (Seehus et al. 2006; Schneider et al. 2011). In an experiment on *A. obtectus*, Lazarević et al. (2013) have found hormetic effect of low concentration of Paraquat (PQ) pro-oxidant agent on base line mortality of females selected for early reproduction and short life-span. Authors hypothesized that early-life hormesis was a consequence of higher egg load and, accordingly, increased content of yolk proteins known to serve as ROS scavenger in PQ challenged insects (Seehus et al. 2006) and be involved in regulation of iron homeostasis and antioxidative defence (Nichol et al. 2002; Pham and Winzerling 2010; Geiser and Winzerling 2012).

In experiment on *Gryllodes sigillatus*, Archer et al. (2012a, 2012b) determined calling effort in males and egg production in females as measures of investment to reproduction, similarly to other experiments with crickets (Hunt et al. 2006; Maklakov et al. 2008; Zajitschek et al. 2009a, 2009b), in order to explore the relationship between oxidative stress and longevity and reproduction. First, they found divergent reproductive strategies between sexes: females maximize reproductive output early in life, while males continually increase calling during ageing. Then, genetic correlations between life history traits and levels of oxidative damage (protein carbonyls) and levels of antioxidative defence (total antioxidative capacity) were determined. As expected, shorter lived sex (females) had higher level of damaged proteins. In both sexes, protein damage was positively correlated with baseline mortality and negatively correlated with life-span and early reproduction. In both sexes, increased damage is followed by increase in antioxidative capacity (positive genetic correlations). However, in contrast to males, late-life protein damage which affected life-span, ageing rate and antioxidative capacity, were not correlated with reproductive effort in females. These results, together with positive genetic correlations across sexes for the level of oxidative damage and antioxidative protection, suggest that sexes differ in managing their redox states and that intra-locus sexual conflict constrained their evolution towards optimal values.

Despite the known importance of mechanisms which repair or remove oxidatively damaged macromolecules, data on sexual dimorphism in these systems are scarce (Østergaard Hansen et al. 2012). Effects of selection for increased longevity and effects of genetic manipulations on the expression of genes for heat shock proteins were mostly studied in males (Kurapati et al. 2000; Tower 2010). Overexpression of Hsp proteins showed some fitness cost (Silberman and Tatar 2000) and effects depended on the tissue and developmental stage (reviewed in Hughes and Reynolds 2005). During ageing

Hsp22-GFP and Hsp70-GFP reporters are up-regulated slightly more in males than females and more in muscle and nervous tissues (Tower et al. 2013). Mutations in *Hsp* genes have detrimental effects more to female than male longevity and diminishes female ability to survive under oxidative stress comparing to control flies (Moskalev et al. 2009). Under control conditions, mutations in genes for transcription factors for heat-shock proteins affects more female than male longevity and some of the mutations showed hormetic effect more often in females than males (Moskalev et al. 2009).

In *D. melanogaster*, activity of superoxide dismutase affects female longevity more than males and positive effects of SOD overexpression was detected in more genetic backgrounds (Spencer et al. 2003). In *Acanthoscelides obtectus* females (longer lived sex) have higher level of catalase activity (Šešlija et al. 1999). In non-virgin *D. simulans*, females (shorter lived sex) have higher production of hydrogen peroxide, lower activity of catalase and mitochondrial SOD, higher proton leak, ADP : O ratio and mtDNA density (Ballard et al. 2007). Since mating is energetically more costly to females, it may be expected that they are more capable to increase metabolic activity in response to reproduction.

3.3.1. The Roles of Mitochondria

Since mitochondria are major generator and target of intracellular ROS, they are suggested to play central role in the ageing process. Age-related decline in mitochondrial function, as well as life-extending effects of mitochondria-targeted genetic and environmental manipulations, demonstrate important role of mitochondria in longevity determination (reviewed in Morrow and Tanguay 2008 and Bratić and Trifunović 2010).

It is well known that functioning of the cell depends on communication between nucleus and mitochondria which, according to direct and retrograde signals, enables adjustment of metabolic reactions with energy requirements. Impairment of mitochondrial function, due to exposure to stress or lack of co-adaptation between nuclear and mitochondrial genomes, decreases ATP production, increases free radical leak and activate regulatory mechanisms which may lead either towards increased expression of respiratory complexes or towards apoptosis (Lane 2011). Outcome will depend on the apoptotic threshold level. Low threshold is related to good mito-nuclear match, low fertility and delayed ageing. As was shown in *Drosophila*, production of reactive oxygen species is lower in long-lived than short-lived lines (Driver and Tawadros 2000) and in females than males (Vina et al. 2003).

Additionally, mitochondria-targeted antioxidants show sex-specific effects on longevity of *D. melanogaster* (Magwere et al. 2006; Krementsova et al. 2012)

It has been suggested that maternal inheritance of mitochondria contributes to sexual dimorphism in ageing and longevity by promoting higher co-adaptation of mitochondrial and nuclear genome in female sex (Zeh and Zeh 2005; Rand 2005; Tower 2006; Wolf and Gemmell 2012). On the other hand, it attenuates the strength of selection on male mitochondrial genome leading to increased mutation load and greater mitochondrial genetic variance for male than female longevity and ageing rate (Gemell et al. 2004; Camus et al. 2012).

In *D. melanogaster*, these mutations affect nuclear gene expression only in males modifying nearly 10% of transcripts mostly in reproductive tissues (Innocenti et al. 2011). As a consequences of asymmetric inheritance of mitochondrial genome, natural and sexually antagonistic selection can maintain mutations highly deleterious to males if they are beneficial to females, or if nuclear genome harbours modifiers which diminish their negative effects. The consequence of the disruption of mito-nuclear interactions may also be revealed as a change in gene expression pattern which lead to increased longevity (Rand et al. 2006).

Several genetic elements, which were mentioned earlier as important elements in regulating life-span, have also been found to have roles in controlling mitochondrial metabolism. Among other functions, transcription factor p53 is involved in regulation of apoptosis and exhibits antagonistic pleiotropic effects on longevity depending on age and sex (Shen and Tower 2010).

Interestingly, sexual dimorphism of p53 effects is under the control of FOXO, while the level of the effects depends on Sir2 gene. Additionally, the mechanisms by which p53 exerts longevity effects might overlap with regulatory pathways which determine feeding behaviour and nutrient utilization.

ACKNOWLEDGMENTS

This work was supported by Ministry of Education, Science and Technological development, grant No. 173007.

References

Adler MI, Cassidy EJ, Fricke C, Bonduriansky R (2013) The lifespan-reproduction trade-off under dietary restriction is sex-specific and context-dependent. *Exp Gerontol* 48: 539–548.

Archer RC (2012) *Food, sex and death: costs of reproduction and the mechanistic basis of ageing.* PhD Dissertation, University of Exeter.

Archer CR, Sakaluk SK, Selman C, Royle NJ, Hunt J (2012a) Oxidative stress and the evolution of sex differences in life span and ageing in the decorated cricket, *Gryllodes sigillatus*. *Evolution* 67: 620–634.

Archer CR, Zajitschek F, Sakaluk SK, Royle NJ, Hunt J (2010b) Sexual selection affects the evolution of lifespan and ageing in the decorated cricket *Gryllodes sigillatus*. *Evolution* 66: 3088–3100.

Arking R, Burde V, Graves K, Hari R, Feldman E, Zeevi A, Soliman S, Saraiya A, Buck S, Vettraino J, Sathrasala K, Wehr N, Levine RL (2000) Forward and reverse selection for longevity in *Drosophila* is characterized by alteration of antioxidant gene expression and oxidative damage patterns. *Exp Gerontol* 35: 167–185.

Arking R, Buck S, Novoseltev VN, Hwangbo D-S, Lane M (2002) Genomic plasticity, energy allocations, and the extended longevity phenotypes of *Drosophila*. *Ageing Res Rev* 1: 209–228.

Bacigalupe LD, Crudington HS, Hunter F, Moore AJ, Snook RR (2007) Sexual conflict does not drive reproductive isolation in experimental populations of *Drosophila pseudoobscura*. *J Evol Biol* 20:1763-1771.

Bahadorani S, Bahadorani P, Phillips JP, Hilliker AJ (2008) The effects of vitamin supplementation on *Drosophila* life span under normoxia and under oxidative stress. *J Gerontol: Biol Sci* 63: 35–42.

Bakkali F, Averbeck S, Averbeck D, Idaomar M (2008) Biological effects of essential oils – A review. *Food Chem Toxicol* 46: 446–475.

Baker DA, Meadows LA, Wang J, Dow JAT, Russell S (2007) Variable sexually dimorphic gene expression in laboratory strains of *Drosophila melanogaster*. *BMC Genomics* 8: 454.

Ballard JWO, Melvin RG, Miller JT, Katewa SD (2007) Sex differences in survival and mitochondrial bioenergetics during aging in *Drosophila*. *Aging Cell* 6: 699–708.

Banerjee KK, Ayyub C, Ali SZ, Mandot V, Prasad NG, Kolthur-Seetharam U (2012) dSir2 in the adult fat body, but not in muscles, regulates life span in a diet-dependent manner. *Cell Rep* 2: 1485-1491.

Bateman PW, Ferguson JWH, Yetman CA (2006) Courtship and copulation, but not ejaculates, reduce longevity of female field crickets (Gryllus bimaculatus). *J Zool* 268: 314–346.

Bauer JH, Morris SN, Chang C, Flatt T, Wood JG, Helfand SL (2009) dSir2 and Dmp53 interact to mediate aspects of CR-dependent lifespan extension in *D. melanogaster*. *Aging* 1:38-48.

Bayne ACV, Mockett RJ, Orr WC, Sohal RS (2005) Enhanced catabolism of mitochondrial superoxide/hydrogen peroxide and aging in transgenic *Drosophila*. *Biochem J* 391: 277–284.

Bede JC, McNeil JN, Tobe SJ (2007) The role of neuropeptides in caterpillar nutritional ecology. *Peptides* 28: 185-196.

Behmer ST (2009) Insect herbivore nutrient regulation. *Ann Rev Entomol* 54: 165–87.

Berg CB, Maklakov AA (2012) Sexes suffer from suboptimal lifespan because of genetic conflict in a seed beetle. *Proc R Soc Lond B* 279: 4296-4302.

Bilde T, Maklakov AA, Meisner K, la Guardia L, Friberg U (2009) Sex differences in the genetic architecture of lifespan in a seed beetle: Extreme inbreeding extends male lifespan. *BMC Evol Biol* 9: 33

Boggs CL (2009) Understanding insect life histories and senescence through a resource allocation lens. *Funct Ecol* 23: 27–37.

Bonduriansky R (2007) The evolution of condition dependent sexual dimorphism. *Am Nat* 167: 9-19.

Bonduriansky R, Maklakov A, Zajitschek F, Brooks R (2008) Sexual selection, sexual conflict and the evolution of ageing and life span. *Funct Ecol* 22: 443–453.

Borash DJ, Rose MR, Mueller LD (2007) Mutation accumulation affects male virility in *Drosophila* selected for later reproduction. *Physiol Biochem Zool* 80: 461-472.

Bratić I, Trifunović A (2013) Mitochondrial energy metabolism and ageing. *Biochim Biophys Acta* 1797: 961–967.

Broughton SJ, Piper MDW, Ikeya T, Bass TM, Jacobson J, Driege Y, Martinez P, Hafen E, Withers DJ, Leevers SJ, Partridge L (2005) Longer lifespan, altered metabolism, and stress resistance in *Drosophila* from ablation of cells making insulin-like ligands. *Proc Natl Acad Sci* USA 102: 3105–3110.

Burger JM, Promislow DE (2004) Sex-specific effects of interventions that extend fly life span. *Sci Aging Knowledge Environ*, pe30.

CamusMF, Clancy DJ, DowlingDK(2012) Mitochondria,maternal inheritance, and male aging. *Curr Biol* 22: 1717–1721

Carey JR, Liedo P, Harshman L, Zhang Y, M H-G, Partridge L, Wang J-L (2002) A mortality cost of virginity at older ages in female Mediterranean fruit flies. *Exp Gerontol* 37: 507-512.

Charlesworth B, Hughes KA (1996) Age-specific inbreeding depression and components of genetic variance in relation to the evolution of senescence. *Proc Natl Acad Sci* USA 93: 6140-6145.

Chippindale AK, Gibbs AG, Sheik M, Yee KJ, Djawdan M, Bradley TJ, Rose MR (1998) Resource acquisition and the evolution of stress resistance in *Drosophila melanogaster*. *Evolution* 52: 1342-1352.

Clancy DJ, Gems D, Harshman LG, Oldham S, Stocker H, Hafen E, Leevers SJ, Partridge L (2001) Extension of life-span by loss of CHICO, a *Drosophila* insulin receptor substrate protein. *Science* 292: 104–106.

Clissold FJ, Tedder BJ, Conigrave AD, Simpson SJ (2010) The gastrointestinal tract as a nutrient-balancing organ. *Proc R Soc* 277B: 1751-1759.

Collins AM, Williams V, Evans JD (2004) Sperm storage and antioxidative enzyme expression in the honey bee, *Apis mellifera*. *Insect Molec Biol* 13: 141–146.

Cotter SC, Simpson SJ, Raubenheimer D, Wilson K (2011) Macronutrient balance mediates trade-offs between immune function and life history traits. *Funct Ecol* 25: 186–198.

Cui X, Wang L, Zuo P, Han Z, Fang Z, Li W, Liu J (2004) D-Galactose-caused life shortening in *Drosophila melanogaster* and *Musca domestica* is associated with oxidative stress. *Biogerontology* 5: 317–325.

De Loof A (2011) Longevity and aging in insects: Is reproduction costly; cheap; beneficial or irrelevant? A critical evaluation of the ''trade-off'' concept. *J Insect Physiol* 57: 1–11.

Demary KC, Lewis SM (2007) Male reproductive allocation in fireflies (*Photinus* spp.). *Invertebr Biol* 126: 74–80.

Demirović D, Rattan SIS (2013) Establishing cellular stress response profiles as biomarkers of homeodynamics, health and hormesis. *Exp Gerontol* 48: 94–98.

De Luca M, Roshina NV, Geiger-Thornsberry GL, Lyman RF, Pasyukova EG, Mackay TFC (2003) Dopa decarboxylase (Ddc) affects variation in *Drosophila* longevity. *Nat Genet* 34: 429–33.

De Luca M, Yi N, Allison DB, Leips J, Ruden DM (2005) Mapping quantitative trait loci affecting variation in *Drosophila* triacylglycerol storage. *Obesity Res* 13: 1596-1605.

Diaz-Albiter HD, Mitford R, Fernando A. Genta FA, Sant'Anna MRV, Dillon RJ (2011) Reactive oxygen species scavenging by catalase is important for female Lutzomyia longipalpis fecundity and mortality. *PLoS ONE* 6(3): e17486.

Djawdan M, Chippindale AK, Rose MR, Bradley TJ (1998) Metabolic reserves and evolved stress resistance in *Drosophila melanogaster. Physiol Zool* 71: 584–594.

Djawdan M, Sugiyama TT, Schlaeger LK, Bradley TJ, Rose MR (1996) Metabolic aspects of the trade-off between fecundity and longevity in *Drosophila melanogaster. Physiol Zool* 69: 1176–1195.

Dowling DK, Simmons LW (2009) Reactive oxygen species as universal constraints in life-history evolution. *Proc R Soc* 276B: 1737–1745.

Driver C, Tawadros N (2000) Cytoplasmic genomes that confer additional longevity in *Drosophila melanogaster. Biogerontology* 1: 255–260.

Dubiley TA, Rushkevich YE, Koshel NM, Voitenko VP, Vaiserman AM (2011) Life span extension in *Drosophila melanogaster* induced by morphine. *Biogerontology* 12: 179–184.

Edvardsson M (2007) Female *Callosobruchus maculatus* mate when they are thirsty: resource rich ejaculates as mating effort in a beetle. *Anim Beh* 74: 183-188.

Ernst IMA, Pallauf K, Bendall JK, Paulsen L, Nikolai S, Huebbe P, Roeder T, Rimbach G (2013) Vitamin E supplementation and lifespan in model organisms. *Ageing Res Rev* 12: 365– 375.

Fedorka KM and Zuk M (2005) Sexual conflict and female immune suppression in the cricket, *Allonemobious socius. J Evol Biol* 18: 1515–1522.

Flatt T, Kawecki TJ (2007) Juvenile hormone as a regulator of the trade-off between reproduction and life span in *Drosophila melanogaster. Evolution* 61: 1980–1991.

Flatt T, Tu M-P, Tatar M (2005) Hormonal pleiotropy and the juvenile hormone regulation of *Drosophila* development and life history. *BioEssays* 27: 999–1010.

Fox CW, Czesak ME, Wallin WG (2004) Complex genetic architecture of population differences in adult lifespan of a beetle: nonadditive inheritance, gender differences, body size and a large maternal effect. *J Evol Biol* 17: 1007-1017.

Fox CW, Dublin L, Pollitt SJ (2003) Gender differences in lifespan and mortality rates in two seed beetle species. *Funct Ecol* 17: 619-626.

Fox CW, Scheibly KL, Wallin WG, Hitchcock Lj, Stillwell RC, Smith BP (2006) The genetic architecture of life span and mortality rates: geneder and species differences in inbreeding load of two seed-feeding beetles. *Genetics* 174: 763-773.

Fox CW, Stillwell RC (2009) Environmental effects on sex differences in the genetic load for adult lifespan in a seed-feeding beetle. *Heredity* 103: 62–72.

Freitas DRJ, Rosa RM, Moraes J, Campos E, Logullo C, Da Silva Vaz Jr. I, Masuda A (2007) Relationship between glutathione S-transferase, catalase, oxygen consumption, lipid peroxidation and oxidative stress in eggs and larvae of *Boophilus microplus* (Acarina: Ixodidae). *Comp Biochem Physiol* 146A: 688–694.

Fu X, South A, Lewis SM (2012) Sexual dimorphism, mating systems, and nuptial gifts in two Asian fireflies (Coleoptera: Lampyridae). *J Insect Physiol* 58: 1485–1492.

Geiser DL, Winzerling JJ (2012) Insect transferrins: multifunctional proteins. *Biochim Biophys Acta* 1820: 437–451.

Gibbs AG, Chippindale AK, Rose MR (1997) Physiological mechanisms of evolved desiccation resistance in *Drosophila melanogaster*. J Exp Biol 200: 1821–1832.

Gong Y, Thompson JN, Woodruff RC (2006) Effect of deleterious mutations on life span in Drosophila melanogaster. *J Gerontol Series A Biol Sci Med Sci* 61: 1246-1252.

Gotthard K, Nylin S, Wiklund C (2000) Mating opportunity and the evolution of sex-specific mortality rates in a butterfly. *Oecologia* 122: 36-43.

Gruntenko NE, Rauschenbach IYu (2008) Interplay of JH, 20E and biogenic amines under normal and stress conditions and its effect on reproduction. *J of Insect Physiol* 54: 902– 908.

Haenold R, Wassef DM, Heinemann SH and Hoshi T (2005) Oxidative damage, aging and anti-aging strategies. *Age* 27: 183–199.

Hall MD, Bussiere LF, Brooks R (2009) Diet-dependent female evolutionary influence male lifespan in a nuptial feeding insect. *J Evol Biol* 22:873-881.

Hallsson LR, Björklund M (2012) Sex-specific genetic variances in life-history and morphological traits of the seed beetle *Callosobruchus maculatus*. *Ecol Evol* 2: 128-138.

Hamilton WD (1966) The moulding of senescence by natural selection. *J Theor Biol* 12: 12-45.

Harshman LG (1999) Investigation of the endocrine system in extended longevity lines of *Drosophila melanogaster*. *Exp Gerontol* 34: 997–1006.

Harshman LG, Moore KM,. Sty MA, Magwire MM (1999) Stress resistance and longevity in selected lines of *Drosophila melanogaster. Neurobiol Aging* 20: 521–529.

Hoffmann J, Romey R, Fink C, Yong L, Roeder T (2013) Overexpression of *Sir2* in the adult fat body is sufficient to extend lifespan of male and female *Drosophila*. *Aging* 5: 315-327.

Holland B, Rice WR (1999) Experimental removal of sexual selection reverses intersexual antagonostic coevolution and removes a rešroductive load. *Proc Natl Acad Sci*, USA 96: 5083-5088.

Hosken DJ, Garner TWJ, Ward PI (2001) Sexual conflict selects for male and female reproductive characters. *Curr Biol* 11: 489-493.

Hughes KA, Alipaz JA, Drnevich JM, Reynolds RM (2002) A test of evolutionary theories of aging. *Proc Natl Acad Sci*, USA99: 14286-14291.

Hughes KA, Reynolds RM (2005) Evolutionary and mechanistic theories of aging. *Ann Rev Entomol* 50: 421-445.

Hunt J, Jennions MD, Spyrou N, Brooks R (2006) Artificial selection on male longevity influences age-dependent reproductive effort in the black field cricket *Teleogryllus commodus*. *Am Nat* 168: E72-E86.

Hwangbo DS, Gershman B, Tu MP, Palmer M, Tatar M (2004) *Drosophila* dFOXO controls lifespan and regulates insulin signalling in brain and fat body. *Nature* 429: 562–566.

Iliadi KG, Natalia N. Iliadi NN, Boulianne GL (2009) Regulation of *Drosophila* life-span: Effect of genetic background, sex, mating and social status. *Exp Gerontol* 44: 546–553.

Innocenti P, Morrow EH, Dowling DK (2011) Experimental evidence supports a sex-specific selective sieve in mitochondrial genome evolution. *Science* 332: 845-848.

Isaksson C, Sheldon BC, Uller T (2011) The challenges of integrating oxidative stress into life-history biology. *BioScience* 61: 194–202.

Ishimoto H, Sakaia T, Kitamoto T (2009) Ecdysone signaling regulates the formation of long-term courtship memory in adult *Drosophila melanogaster*. *Proc Natl Acad Sci* 106: 6381–6386.

Ivy TM, Sakaluk SK (2005) Polyandry promotes enhanced offspring survival in decorated crickets. *Evolution* 59: 152–159.

Jaureguy LM, Etges WJ (2007) Assessing patterns of senescence in *Drosophila mojavensis* reared on different host cacti. *Evol Ecol Res* 9: 1–17.

Jervis MA, Boggs CL, Ferns PN (2005) Egg maturation strategy and its associated trade-offs: a synthesis focusing on Lepidoptera. *Ecol Entomol* 30: 359–375.

Jin W, Riley RM, Wolfinger RD, White KP, Passador-Gurgel G, Gibson G (2001) The contributions of sex, genotype and age to transcriptional variance in *Drosophila melanogaster*. *Nat Genet* 29: 389-395.

Jünger M, Rintelen F, Stocker H, Wasserman J, Végh M, Radimerski T, Greenberg M, Hafen E (2003) The *Drosophila* Forkhead transcription factor FOXO mediates the reduction in cell number associated with reduced insulin signaling. *J Biol* 2(3):20.

Kirkwood TBL (1977) Evolution of ageing. *Nature* 270: 301-304.

Krementsova AV, Roshina NV, Tsybul'ko EA, Rybina OY, Symonenko AV, Pasyukova EG (2012) Reproducible effects of the mitochondria-targeted plastoquinone derivative SkQ1 on *Drosophila melanogaster* lifespan under different experimental scenarios. *Biogerontology* 13: 595–607.

Kurama T, Kurata S, Natori S (1995) Molecular characterization of an insect transferrin and its selective incorporation into eggs during oogenesis. *Eur J Biochem* 228: 229–235.

Kurapati R, Passananti HB, Rose MR, Tower J (2000) Increased *hsp22* RNA levels in *Drosophila* lines genetically selected for increased longevity. *J Gerontol: Biol Sci* 55A: B552–B559.

Lande R (1980) Sexual dimorphism, sexual selection and adaptation in polygenic characters. *Evolution* 38:292-305.

Lane N (2011) Mitonuclear match: Optimizing fitness and fertility over generations drives ageing within generations. *BioEssays* 33: 860–869.

Lazarević J (1994) The effect of host plant on diugestive enyzmes and variation of fitness components in the gypsy moth *Lymantria dispar* L. MS Thesis, University of Belgrade.

Lazarević J, Đorđević M, Stojković B, Tucić N (2013) Resistance to prooxidant agent paraquat in the short- and long-lived lines of the seed beetle (*Acanthoscelides obtectus*). *Biogerontology* (2013) 14:141–152.

Lazarević J, Tucić N, Šešlija Jovanović D, Večeřa J, Kodrík D (2012) The effects of selection for early and late reproduction on metabolite pools in *Acanthoscelides obtectus* Say. *Insect Sci* 19: 303–314.

Le Bourg É (2001) Oxidative stress, aging and longevity in *Drosophila melanogaster*. *FEBS Letters* 498: 183-186.

Le Bourg É (2007a) Hormetic effects of repeated exposures to cold at young age on longevity, aging and resistance to heat or cold shocks in *Drosophila melanogaster*. *Biogerontology* 8: 431–444.

Le Bourg É (2007b) Hormetic effects on longevity of hydrogen peroxide in *Drosophila melanogaster* flies living on a poorly nutritious medium. *Biogerontology* 8: 327–344.

Le Bourg É (2012) Combined effects of two mild stresses (cold and hypergravity) on longevity, behavioral aging, and resistance to severe stresses in *Drosophila melanogaster*. *Biogerontology* 13: 313–328.

Le Bourg É, Massou I, Vanessa Gobert V (2009) Cold stress increases resistance to fungal infection throughout life in *Drosophila melanogaster*. *Biogerontology* 10: 613–625.

Lee KP, Simpson SJ, Clissold FJ, Brooks R, Ballard JWO, Taylor PW, Soran N, Raubenheimer D (2008) Lifespan and reproduction in *Drosophila*: New insights from nutritional geometry. *Proc Natal Acad Sci* 105: 2498-2503.

Lesser KJ, Paiusi IC, Leips J (2006) Naturally occurring genetic variation in the age-specific immune response of *Drosophila melanogaster*. *Aging Cell* 5: 293-295.

Lewis Z, Wedell N, Hunt J (2011) Evidence for strong intralocus sexual conflict in the Indian meal moth, *Plodia interpunctella*. *Evolution* 65: 2085-2097.

Loder D (2006) Effect of hsp70 overexpression on life span and stress resistance in *Drosophila melanogaster*. Doctoral dissertation, Quin's University, Kingston, ON.

Magwere T, Chapman T, Partridge L (2004) Sex differences in the effect of dietary restriction on lifespan and mortality rates in female and male *Drosophila melanogaster*. *J Gerontol A: Biol Sci Med Sci* 59: B3–B9.

Magwere T, West M, Riyahi K, Murphy MP, Smith RAJ and Partridge L (2006) The effects of exogenous antioxidants on lifespan and oxidative stress resistance in *Drosophila melanogaster*. *Mech Ageing Dev* 127: 356–370.

Magwire MM, Yamamoto A, Carbone MA, Roshina NV, Symonenko AV, Pasyukova EG, Morozova TV, Mackay TFC (2010) Quantitative and molecular genetic analyses of mutations increasing Drosophila life span. *PLoS Genet* 6(7): e1001037.

Mair W, Sgrò CM, Johnson AP, Chapman T, Partridge L (2004) Lifespan extension by dietary restriction in female *Drosophila melanogaster* is not caused by a reduction in vitellogenesis or ovarian activity. *Exp Gerontol* 39: 1011–1019.

Maklakov AA, Bonduriansky R, Brooks RC (2009a) Sex differences, sexual selection, and ageing: An experimental evolution approach. *Evolution* 63: 2491-2503.

Maklakov AA, Hall MD, Simpson SJ, Dessmann J, Clissold FJ, Zajitschek F, Lailvaux SP, Raubenheimer D, Bonduriansky R, Brooks RC (2009b) Sex differences in nutrient-dependent reproductive ageing. *Aging Cell* 8:324–330.

Maklakov AA, Cayetano L, Brooks RC, Bonduriansky R (2010) The role of life-history selection and sexual selection in the adaptive evolution of mating behavior in a beetle. *Evolution* 64: 1273-1282.

Maklakov AA, Fricke C (2009) Sexual selection did not contribute to the evolution of male lifespan under curtailed age at reproduction in a seed beetle. *Ecol Entomol* 34: 638_643.

Maklakov AA, Fricke C, Arnqvist G (2007) Sexual selection affects lifespan and ageing in the seed beetle. *Aging Cell* 6: 739-744.

Maklakov AA, Kremer N, Arnqvist G (2005) Adaptive male effects on female ageing in seed beetles. *Proc R Soc* 272: 2485-2489.

Maklakov AA, Lummaa V (2013) Evolution of sex differences in lifespan and aging: causee and constraints. *BioEssays* Doi 10.1002/bies.201300021.

Maklakov AA, Simpson SJ, Zajitschek F, Hall MD, Dessmann J, Clissold F, Raubenheimer D, Bonduriansky R, Brooks RC (2008) Sex-specific fitness effects of nutrient intake on reproduction and lifespan. *Curr Biol 18*: 1062–1066.

Mank JE, Wedell N, Hosken DJ (2013) Polyandry and sex-specific gene expression. *Philos Trans R Soc Lond B Biol Sci* 368: 20120047.

Martin OY, Hosken DJ (2003) Costs and benefits of evolving under experimentally enforced polyandry or monogamy. *Evolution* 57: 2765-2772.

Matzkin LM, Mutsaka K, Johnson S, Markow TA (2009) Metabolic pools deffer among ecologically diverse *Drosophila* species. *J Insect Physiol* 55: 1145-1150.

May RC (2007) Gender, immunity and the regulation of longevity. *BioEssays* 29: 795-802.

Mayr E (1961) Cause and effect in biology. *Science* 134: 1501-1506.

Medawar PB (1952) An unsolve problem of biology. HK Lewis, London.

Melvin RG, Ballard WO (2011) Females with a mutation in a nuclear-encoded mitochondrial protein pay a higher cost of survival than do males in *Drosophila. J Geront A Biol Sci Med Sci* 66A: 765-770.

Min K-J, Tatar M (2006) Restriction of amino acids extends lifespan in *Drosophila melanogaster*. *Mech Ageing Dev* 127: 643–646.

Minois N (2000) Longevity and aging: beneficial effects of exposure to mild stress. *Biogerontology* 1: 15-29.

Minois N (2001) Resistance to stress as a function of age in transgenic *Drosophila melanogaster* overexpressing Hsp70. *J Insect Physiol* 47: 1007–1012.

Miwa S, Riyahi K, Partridge L, Brand MD (2004) Lack of correlation between mitochondrial reactive oxygen species production and life span in *Drosophila*. *Ann NY Acad Sci* 1019: 388-391.

Mockett RJ, Orr WC, Rahmandar JJ, Sohal BH and Sohal RS (2001) Antioxidant status and stress resistance in long- and short-lived lines of *Drosophila melanogaster*. *Exp Gerontol* 36: 441-463.

Mockett RJ, Sohal RS (2006) Temperature-dependent trade-offs between longevity and fertility in the *Drosophila* mutant, *methuselah*. *Exp Gerontol* 41: 566–573.

Moczek AP (2009) Phenotypic plasticity and the origin of diversity: a case study on horned beetles, pp 81-134. In *Phenotypic Plasticity of Insects* (ed. by DW Whitman and T.N. Ananthakrishnan), Science Publishers, Enfield, NH, USA.

Monaghan P, Metcalfe NB, Torres R (2009) Oxidative stress as a mediator of life history trade-offs: mechanisms, measurements and interpretation. *Ecol Letters* 12: 75–92.

Morrow EH, Arnqvist G (2003) Costly traumatic insemination and a female counter-adaptation in bed bugs. *Proc R Soc Lond* B 270B: 2377–2381.

Morrow G, Tanguay RM (2008) Mitochondria and ageing in *Drosophila*. *Biotechnol. J* 3: 728–739.

Moskalev A, Shaposhnikov M, Turysheva E (2009) Life span alteration after irradiation in *Drosophila melanogaster* strains with mutations of Hsf and Hsps. *Biogerontology* 10: 3–11.

Moskalev AA, Plyusnina EN, Shaposhnikov MV (2011) Radiation hormesis and radioadaptive response in *Drosophila melanogaster* flies with different genetic backgrounds: the role of cellular stress-resistance mechanisms. *Biogerontology* 12:253–263.

Mourikis P, Hurlbut GD, Artavanis-Tsakonas S (2006) Enigma, a mitochondrial protein affecting lifespan and oxidative stress response in *Drosophila*. *Proc Natl Acad Sci* USA 103: 1307-1312.

Nakagawa S, Lagisz M, Hector KL, Spencer HG (2012) Comparative and metaanalytic insights into life extension via dietary restriction. *Aging Cell* 11: 401–409.

Nichol H, Law JH, Winzerling JJ (2002) Iron metabolism in insects. *Annu Rev Entomol* 47: 535–559.

Norry FM, Loeschcke VR (2002) Longevity and resistance to cold stress in cold-stress selected lines and their controls in *Drosophila melanogaster*. *J Evol Biol* 15: 775–783.

Nuzhdin SV, Pasyukova EG, Dilda CL, Zeng Z-B, Mackay TFC (1997) Sex-specific quantitative trait loci affecting longevity in *Drosophila melanogaster*. *Proc Natl Acad Sci USA* 94: 9734–9739.

Orr WC, Radyuk SN, Prabhudesai L, Toroser D, Benes JJ, Luchak JM, Mockett RJ, Rebrin I, Hubbard JG, Sohal RS (2005) Overexpression of glutamate-cysteine ligase extends life span in *Drosophila melanogaster*. *J Biol Chem* 280: 37331–37338.

Østergaard Hansen T, Sarup P, Loeschcke V, Rattan SIS (2012) Age-related and sex-specific differences in proteasome activity in individual *Drosophila* flies from wild type, longevity-selected and stress resistant strains. *Biogerontology* 13: 429–438.

Parker GA (1979) Sexual selection and sexual conflict, pp. 123-166. In: *Sexual Selection and Reproductive Competition in Insects*, (ed. by MS Blum and NA Blum). Academic Press: New York.

Partridge L, Fowler K (1992) Direct and correlated responses to selection on age at reproduction in *Drosophila melanogaster*. *Evolution* 46: 76-91.

Partridge L, Piper MD, Mair W (2005) Dietary restriction in *Drosophila*. *Mech Ageing Dev* 126: 938–950.

Peretz G, Bakhrat A, Abdu U (2007) Expression of the *Drosophila melanogaster* GADD45 homolog (CG11086) affects egg asymmetric development that is mediated by the c-Jun N-terminal kinase pathway. *Genetics* 177:1691–1702.

Pham DQD, Winzerling JJ (2010) Insect ferritins: typical or atypical? *Biochim Biophys Acta* 1800: 824–833.

Piper MDW, Partridge L (2007) Dietary restriction in drosophila: delayed aging or experimental artefact? *PLoS Genetics* 3: e57.

Piper MDW, Partridge L, Raubenheimer D, Simpson SJ (2011) Dietary restriction and aging: A unifying perspective. *Cell Metabol* 14:154-160.

Plyusnina EN, Shaposhnikov MV, Moskalev AA (2011) Increase of *Drosophila melanogaster* lifespan due to D-GADD45 overexpression in the nervous system. *Biogerontology* 12:211–226.

Poon PC, Kuo T-H, Linford NJ, Roman G, Pletcher SD (2010) Carbon Dioxide Sensing Modulates Lifespan and Physiology in *Drosophila. PLoS Biol* 8(4): e1000356.

Promislow DEL and Haselkorn TS (2002) Age-specific metabolic rates and mortality rates in the genus *Drosophila. Aging Cell* 1: 66–74.

Pujol-Lereis LM, Rabossi A, Quesada-Allué LA (2012) Lipid profiles as indicators of functional senescence in the medfly. *Exp Gerontol* 47: 465–472.

Rand DM (2005) Mitochondrial genetics of aging: intergenomic conflict resolution. *Sci. Aging Knowledge Environ* re5.

Rand DM, Fry A, Sheldahl L (2006) Nuclear–mitochondrial epistasis and *Drosophila* aging: introgression of *Drosophila simulans* mtDNA modifies longevity in *D. melanogaster* nuclear backgrounds. *Genetics* 172: 329–341.

Rattan SIS (2012) Biogerontology: from here to where? The Lord Cohen Medal Lecture—2011. *Biogerontology* 13: 83–91.

Reed DH, Braynt EH (2000) The evolution of senescence under curtailed life span in laboratory populations of *Musca domestica* (the housfly). *Heredity* 85: 115-121.

Reis T, Van Gilst MR, Hariharan IK (2010) A buoyancybased screen of *Drosophila* larvae for fat-storage mutants reveals a role for Sir2 in coupling fat storage to nutrient availability. *PLoS Genet* 6: e1001206.

Reynolds RM, Temiyasathit S, Reedy MM, Ruedi EA, Drnevich JM, Leips J, Hughes KA (2007) Age specificity of inbreeding load in *Drosophila melanogaster* and implications for the evolution of late-life mortality plateaus. *Genetics* 177: 587-595.

Rogina B, Helfand SL (2004) Sir2 mediates longevity in the fly through a pathway related to calorie restriction. *Proc Natl Acad Sci* 101: 15998–16003.

Rolff J, Siva-Jothy MT (2002) Copulation corrupts immunity: a mechanism for a cost of mating in insects. *Proc* Natl Acad Sci USA 99: 9916–9918.

Rönn J, Katvala M, Arnqvist G (2006) The costs of mating and egg production in *Callosobruchus* seed beetles. *Animal Behav* 72: 335-342.

Rose MR (1984) Laboratory evolution of postponed senescence in *Drosophila melanogaster. Evolution* 38: 1004-1010.

Rose MR, Charlesworth B (1980) A test of evolutionary theories of senescence. *Nature* 287: 141-142.

Rose MR, Drapeau MD, Yazdi PG, Shah KH, Moise DB, Thakar RR, Rauser CL, Mueller LD (2002) Evolution of late-life mortality in *Drosophila melanogaster*. *Evolution* 56: 1982-1991.

Rose MR, Passananti HB, Chippindale AK, Phelan JP, Matos M, Teotonio H, Mueller LD (2005) The effects of evolution are local: evidence from experimental evolution in *Drosophila*. *Integr Comp Biol* 45: 486-491.

Ruiz M, Sanchez D, Canal I, Acebes A, Ganfornina MD (2011) Sex-dependent modulation of longevity by two *Drosophila* homologues of human Apolipoprotein D, GLaz and NLaz. *Exp Gerontol* 46: 579–589.

Sanz A, Fernández-Ayala DJM, Stefanatos RKA, Jacobs HT (2010) Mitochondrial ROS production correlates with, but does not directly regulate lifespan in drosophila. *Aging* 2: 200-223.

Sarup P, Loeschcke V (2011) Life extension and the position of the hormetic zone depends on sex and genetic background in *Drosophila melanogaster*. *Biogerontology* 12:109–117.

Savalli VM, Fox CW (1999) The effect of male size, age, and mating behaviour on sexual selection in the beetles *Callosobruchus maculatus*. *Ecol Ethol Evol* 11: 49-60.

Scannapieco AC, Sambucetti P, Norry FM (2009) Direct and correlated responses to selection for longevity in *Drosophila buzzatii*. *Biol J Linn Soc* 97: 738–748.48

Scannapieco AC, Sørensen JG, Loeschcke V, Norry FM (2007) Heat-induced hormesis in longevity of two sibling *Drosophila* species. *Biogerontology* 8:315–325

Schneider SA, Schrader C, Wagner AE, Boesch-Saadatmandi C, Liebig J, Rimbach G, Roeder T (2011) Stress resistance and longevity are not directly linked to levels of enzymatic antioxidants in the Ponerine Ant *Harpegnathos saltator*. *PLoS ONE* 6(1): e14601. doi:10.1371/ journal.pone.0014601.

Seehuus S-C, Norberg K, Gimsa U, Krekling T, Amdam GV (2006) Reproductive protein protects functionally sterile honey bee workers from oxidative stress. *Proc Natl Acad Sci USA* 103: 962-967.

Seto NOL, Hayashi S, and Tener GM (1990) Overexpression of Cu-Zn superoxide dismutase in *Drosophila* does not affect life-span. *Proc Natl Acad Sci* USA 8: 4270-4274.

Shen J, Ford D, Landis GN, Tower J (2009) Identifying sexual differentiation genes that affect *Drosophila* life span. *BMC Geriatrics* 9: 56.

Shen J, Tower J (2010) *Drosophila* foxo acts in males to cause sexual-dimorphism in tissue-specific p53 life span effects. *Exp Gerontol* 45: 97–105.

Shuker DM, Ballantyne GA, Wedell N (2006) Variation in the cost to females of the sexual conflict over mating in the seed bug, *Lygaeus equestris*. *Animal Behav* 72: 313-321.

Silbermann R, Tatar M (2000) Reproductive costs of heat shock protein in transgenic *Drosophila melanogaster*. *Evolution* 54: 2038-2045.

Simmons LW, García-González F (2008) Evolutionary reduction in testes size and competitive fertilization success in response to the experimental removal of sexual selection in dung beetles. *Evolution* 62: 2580–2591.

Simon AF, Shih C, Mack A, Benzer S (2003) Steroid control of longevity in *Drosophila melanogaster*. *Science* 299: 1407–1410.

Simpson SJ, Raubenheimer D (2009) Macronutrient balance and lifespan. *Aging* 1: 875-880.

Slack C, Werz C, Wieser D, Alic N, Foley A, Stocker H, Withers DJ, Thornton JM, Hafen E, Partridge L (2010) Regulation of lifespan, metabolism, and stress Responses by the *Drosophila* SH2B protein, Lnk. *PLoS Genet* 6(3): e1000881.

Sohal RS, Orr WC (2012) The Redox Stress Hypothesis of Aging. *Free Radic Biol Med* 52: 539-555.

Sørensen JG, Kristensen TN, Kristensen KV, Loeschcke V (2007) Sex specific effects of heat induced hormesis in Hsf-deficient *Drosophila melanogaster*. *Exp Gerontol* 42: 1123–1129.

South SH, House CM, Moore AJ, Simpson SJ, Hunt J (2011) Male cockroaches prefer a high carbohydrate diet that makes them more attractive to females: implications for the study of condition dependence. *Evolution* 65: 1594–1606.

Spencer CC, Howell CE, Wright AR, Promislow DEL (2003) Testing an 'aging gene' in long-lived *Drosophila* strains: Increased longevity depends on sex and genetic background. *Aging Cell* 2: 123–130.

Stenesen DS (2011) In pursuit of a molecular fountain of youth: the identification and characterization of lifespan regulators in *Drosophila*. PhD Dissertation, University of Texas.

Stockhoff BA (1993) Ontogenetic change in dietary selection for protein and lipid by gypsy moth larvae. *J Insect Physiol* 39: 677-686.

Stojković B, Savković U (2011) Gender differences in longevity in early and late reproduced lines of the seed beetle. *Arch Biol Sci*, Belgrade 63: 129-136.

Šešlija D, Blagojević D, Spasić M and Tucić N (1999). Activity of superoxide dismutase and catalase in the bean weevil (*Acanthoscelides obtectus*) selected for postponed senescence. *Exp Gerontol* 34: 185-195.

Šešlija D, Marečko I, Tucić N (2008) Sexual selection and senescence, do seed beetle males (*Acanthoscelides obtectus*, Bruchidae, Coleoptera) shape longevity of their mates? *J Zool Syst Evol Res* 46: 323–330.

Šešlija D, Tucić N (2008) The genetic architecture of extended life span in the seed beetle *Acanthoscelide obtectus* (Coleoptera: Bruchidae). *Eur J Entomol* 105: 553-560.

Tanaka Y (1990) Age specificity of inbreeding depression during a life cycle of *Callosobruchus chinensis* (Coleoptera: Bruchidae). *Res Pop Ecol* 32: 329-335.

Tatar M, Bartke A, Antebi A (2003) The endocrine regulation of aging by insulin-like signals. *Science* 299: 1346-1351.

Tatar M, Kopelman A, Epstein D, Tu MP, Yin CM, Garofalo RS (2001) A mutant *Drosophila* insulin homolog that extends life-span and impairs neuroendocrine function. *Science* 292: 107–110.

Toivonen JM, Partridge L (2009) Endocrine regulation of ageing and reproduction in *Drosophila*. *Molec Cellular Endocrinol* 299: 39-50.

Tower J (2006) Sex-specific regulation of aging and apoptosis. *Mech Ageing Dev* 127: 705–718.

Tower J, Arbeitman M (2009) The genetics of gender and life span. *J Biol* 8: 38 (doi:10.1186/jbiol141).

Tower J (2010) Heat shock proteins and *Drosophila* aging. *Exp Gerontol* 46: 355-362.

Tower J, Landis G, Gao R, Luan A, Lee J, Sun Y (2013) Variegated expression of *Hsp22* transgenic reporters indicates cell-specific patterns of aging in *Drosophila* oenocytes. *J Gerontol A Biol Sci Med Sci* doi:10.1093/gerona/glt078.

Tu M-P, Epstein D, Tatar M (2002) The demography of slow aging in male and female *Drosophila* mutant for the insulin-receptor substrate homologue *chico*. *Aging Cell* 1: 75–80

Tucić N, Cvetković D, Milanović D (1988) The genetic variation and covariation among ditness compoents in *Drosophila melanogaster* females and males. *Heredity* 60:55-60.

Tucić N, Gliksman I, Šešlija D, Milanović D, Mikuljanac S, Stojković O (1996) Laboratory evolution of longevity in the bean weevil (*Acanthoscelides obtectus*). *J Evol Biol 9*: 485-503.

Tucić N, Šešlija D, Stanković V (2004) The short-term and long-term effects of parental age in the bean weevil (*Acanthoscelides obtectus*). *Evol Ecol* 18: 187–201.

Vaiserman AM, Koshel NM, Litoshenko AY, Mozzhukina TG, Voitenko VP (2003) Effects of X-irradiation in early ontogenesis on the longevity and amount of the S1 nuclease-sensitive DNA sites in adult *Drosophila melanogaster*. *Biogerontology* 4: 9–14.

Vaiserman AM, Zabuga OG, Kolyada AK, Pisaruk AV, Kozeretska IA (2013) Reciprocal cross differences in *Drosophila melanogaster* longevity: an evidence for non-genomic effects in heterosis phenomenon? *Biogerontology* DOI 10.1007/s10522-013-9419-6.

Vermeulen CJ, Cremers TIFH, Westerink BHC, Van De Zande L, Bijlsma R (2006a) Changes in dopamine levels and locomotor activity in response to selection on virgin life span in *Drosophila melanogaster*. *Mech Ageing Dev* 127: 610-617.

Vermeulen CJ, Van De Zande L, Bijlsma R (2006b) Devlopmental and age-specific effects of selection on divergent virgin life span on fat content and starvation resistance in *Drosophila melanogatster*. *J Insect Physiol* 52: 910-919.

Vermeulen CJ, Loeschcke V (2007) Longevity and the stress response in *Drosophila*. *Exp Gerontol* 42: 153–159.

Vina J, Sastre J, Pallardo F, Borras C (2003) Mitochondrial theory of aging: importance to explain why females live longer than males. *Antioxid Redox Signal* 5: 549–556.

Wagner WE, Kelley RJ, Tucke KR, Harper CJ (2001) Females receive a life-span benefit from male ejaculates in a field cricket. *Evolution* 55: 994–1001.

Wang H-D, Kazemi-Esfarjani P, Benzer S (2004) Multiple-stress analysis for isolation of *Drosophila* longevity genes. *Proc Natl Acad Sci* USA 101: 12610–12615.

Waskar M, Landis GN, Shen J, Curtis C, Tozer K, Abdueva D, Skvortsov D, Tavaré S, Tower J (2009) *Drosophila melanogaster p53*has developmental stage-pecific and sex-specific effects on adult life span indicative of sexual antagonistic pleiotropy. *Aging* 1: 903-936.

Wayne ML, Telonis-Scott M, Bono LM, Harshman L, Kopp A, Nuzhdin SV, McIntyre LM (2007) Simpler mode of inheritance of transcriptional variation in male *Drosophila melanogaster*. *Proc Natl Acad Sci*, USA 104: 18577-18582.

Williams GC (1957) Pleiotropy, natural selection and the evolution of senescence. *Evolution* 11: 398-411.

Wilson RH, Lai CQ, Lyman RF, Mackay TFC (2013) Genomic response to selection for postponed senescence in *Drosophila*. *Mech Ageing Dev* 134: 79–88.

Wit J, Sarup P, Lupsa N, Malte H, Frydenberg J, Loeschcke V (2013) Longevity for free? Increased reproduction with limited trade-offs in Drosophila melanogaster selected for increased life span. *Exp Gerontol* 48: 349–357.

Wolf JN, Gemmell NJ (2012) Mitochondria, maternal inheritance, and asymmetric fitness: why males die younger? *BioEssays* 35: 93–99.

Yee WKW, Sutton KL, Dowling DK (2013) In vivo male fertility is affected by naturally occurring mitochondrial haplotypes. *Curr Biol* 23: R55-56.

Yılmaz M, Özsoy ED, Bozcuk AN (2008) Maternal age effects on longevity in Drosophila melanogaster populations of different origin. *Biogerontology* 9:163–168.

Zajitschek F, Bonduriansky R, Zajitschek SRK, Brooks RC (2009a) Sexual dimorphism in life history: age, survival, and reproduction in male and female fieldcrickets *Teleogryllus commodus* under seminatural conditions. *Am Nat* 173: 792-802.

Zajitschek F, Hunt J, Jennions MD, Hal MD, Brooks RC (2009b) Effects of juvenile and adult diet on ageing and reproductive effort of male and female black field crickets, *Teleogryllus commodus*. *Funct Ecol* 23: 602–611.

Zeh JA, Zeh DW (2005) Maternal inheritance, sexual conflict and the maladapted male. *TRENDS Gen* 21: 281-286.

Zera AJ (2005) Intermediary metabolism and life history trade-offs: lipid metabolism in lines of the Wing-polymorphic cricket, *Gryllus firmus*, selected for flight capability *vs.* early age reproduction. *Integr Comp Biol* 45: 511–524.

Zera AJ, Zhao Z, Kaliseck K (2007) Hormones in the field: Evolutionary endocrinology of juvenile hormone and ecdysteroids in field populations of the Wing-Dimorphic Cricket *Gryllus firmus*. *Physiol Biochem Zool* 80: 592–606.

In: Sexual Selection
Editors: R. Geldani and M. Davin
ISBN: 978-1-62808-805-2

Chapter 2

EVIDENCE OF NATURAL AND SEXUAL SELECTION SHAPING THE SIZE OF NUPTIAL GIFTS AMONG A SINGLE BUSH-CRICKET GENUS (*POECILIMON*; TETTIGONIIDAE): AN ANALYSIS OF SPERM TRANSFER PATTERNS

J. McCartney [1,2,*]***, M. A. Potter***[1]***, A. W. Robertson***[1]***,***
K-G. Heller[2] ***and D. T. Gwynne***[3]
[1]Ecology Group, Institute of Agriculture and Environment, Massey University, Palmerston North, New Zealand
[2]Friedrich Alexander Universität, Institute of Biology, Erlangen – Nürnberg, Germany
[3]Biology Department, University of Toronto in Mississauga, Mississauga, Canada

ABSTRACT

During mating, male bush-crickets transfer a complex spermatophore to the female. The spermatophore is comprised of a large nuptial gift which the female consumes while the sperm from the ejaculate-

* Corresponding author (Email: J.McCartney@massey.ac.nz).

containing ampulla are transferred into her. Two main functions of the nuptial gift have been proposed: the ejaculate protection hypothesis and the parental investment hypothesis. The former, founded on sexual selection theory, predicts that the time to consume the gift is no longer than necessary to allow for full ejaculate transfer. The latter maintains that gift nutrients increase the fitness or quantity of offspring and hence the gift is likely to be larger than is necessary for complete sperm transfer. With an aim to better understanding the primary function of nuptial gifts, we examined sperm transfer data from field populations of five *Poecilimon* bush-cricket taxa with varying spermatophore sizes. In the species with the largest spermatophore, the gift was four times larger than necessary to allow for complete sperm transfer and is thus likely to function as paternal investment. Species with medium and small gifts were respectively sufficient and insufficient to allow complete sperm transfer and are likely to represent, to various degrees, ejaculate protection. We also found that species that produce larger spermatophores transfer greater proportions of available sperm than species producing smaller spermatophores, and thus achieve higher paternal assurance.

Keywords: Ejaculate protection, mating effort, paternal investment, spermatophore size, sperm transfer, sperm competition

INTRODUCTION

Nuptial feeding has been observed in several insect taxa (Thornhill and Alcock, 1983; Vahed, 1998). Male bush-crickets (Tettigoniidae) transfer a substantial, and often costly, spermatophore to the females for consumption during mating (Wedell, 1994a, 1994b; Vahed, 2007a). The spermatophore consists of an ampulla that contains the sperm, and a spermatophylax that, in most species, is a large gelatinous mass. The female first eats the spermatophylax and then eats the smaller ampulla along with any remaining sperm and seminal fluid (Bowen et al., 1984). There is debate over the selective pressures that maintain nuptial gift size in bush-crickets (for reviews see Thornhill and Alcock, 1983; Simmons and Parker, 1989; Vahed, 1998; Gwynne, 2001; Vahed, 2007; Gwynne, 2008; McCartney, et al 2008, 2010, 2012). Despite recent discussions concerning the effect of sexual conflict on nuptial gift size (e.g. Vahed, 2007b; Gwynne, 2008; Lehman, 2012) two hypotheses remain central to understanding the role of gift size: the ejaculate protection hypothesis and the parental investment hypothesis.

The ejaculate protection hypothesis argues that the nuptial gift is sexually selected; it increases fertilisation success by diverting the female away from the sperm ampulla while maximum insemination is achieved (Gerhard, 1913; Boldyrev, 1915; Gwynne, 1984; Sakaluk and Eggert, 1996; Vahed and Gilbert, 1996; Simmons, 2001). The parental investment hypothesis proposes that the function of the nuptial gift is derived from its nutritive value and that these nutrients are passed into the donating males' offspring; the gift is thus under natural selection to increase the quality and/or the quantity of the male's offspring (Trivers, 1972; Thornhill, 1976; Gwynne, 1986, 1988a, 1988b, 1990; Reinhold, 1999).

The ejaculate protection and paternal investment hypotheses are not mutually exclusive (Quinn and Sakaluk, 1986) and present research focuses on the relative importance of the two hypotheses in different taxa. It is likely that the spermatophylax evolved through sexual selection for ejaculate protection in bush-crickets (Gwynne, 1986, 1990, 1997, 2001), but there is evidence that both functions can be involved in the maintenance of spermatophylax size in various tettigoniid species (for reviews see Vahed, 1998; Gwynne, 2001; McCartney et al. 2008).

Nuptial gifts that function to protect the ejaculate are predicted to be smaller, less nutritious, and of a size that co-varies with either sperm number and/or ampulla size and should be no larger than necessary to allow for complete insemination (Reinhold and Heller, 1993; Wedell, 1993a, 1994a, 1994b; Heller and Reinhold, 1994; Vahed and Gilbert, 1996). Nuptial gifts that are influenced by paternal investment are likely to be large, nutritious (Wedell, 1994a, 1994b), and take longer to consume than it takes to transfer a full complement of sperm (Wedell, 1994b). While it can be relatively simple to test the prediction of the ejaculate protection hypothesis, at least three further criteria underpin paternal investment in nuptial-gift-bearing species and are needed to distinguish it from the ejaculate protection hypothesis: 1) the degree of last-male mating advantage; 2) the time that it takes for the nutrients of the spermatophylax to directly affect the donating males' offspring; and 3) the relationship between female mating interval and egg laying interval (see Vahed, 1998 and references cited therein).

The ejaculate protection hypothesis is supported by comparative studies across taxa showing positive correlations between spermatophylax size and ampulla mass or sperm number (Wedell, 1993a; Vahed and Gilbert, 1996; McCartney et al., 2008, 2012), as well as studies within species showing that the size of the nuptial gift or the consumption time of the gift is roughly similar to the time that it takes for the majority of sperm to transfer into the female (e.g. Wedell and Arak, 1989; Wedell, 1991; Reinhold and Heller, 1993; Heller and Reinhold, 1994; Vahed, 1994; Simmons, 1995a). Evidence of paternal investment has also been observed in some species (Gwynne et al., 1984; Gwynne, 1986, 1988a, 1988b; Simmons, 1990; Wedell, 1994a, 1994b; Simmons et al., 1999; Reinhold, 1999), yet almost all insect species studied thus far, including those with properties of paternal investment, have nuptial gifts (or nuptial gift consumption times) that approximate the size necessary for complete sperm transfer (Heller and Reinhold, 1994; Simmons, 1995a; Simmons and Gwynne, 1991; Vahed, 1994), and are therefore likely to be maintained primarily through sexual selection via the ejaculate protection hypothesis (Vahed, 1998).

Diverse examples of this rule can be found in Mecoptera as prey and salivary masses, Diptera as nuptial prey and regurgitated food, Coleoptera and Zoraptera as cephalic gland secretions, and other Orthoptera, as hind-wing and glandular secretion feeding (Vahed, 1998 and references cited therein).

Possibly the only exception is *Requena verticalis*, initially reported to have a spermatophylax twice as large as necessary to allow for complete sperm transfer of the ampulla (Gwynne et al., 1984; Gwynne, 1986, 1988*b*). However, further research on this species (Simmons, 1995*a*, 1995*b*; Simmons et al., 1999) and different interpretations of what constitutes 'complete' sperm transfer (Vahed, 1994, 1998; Simmons, 1995*a*) suggest that complete sperm transfer may not be achieved until close to, or even after gift consumption (Vahed, 1998).

Additionally, males have a substantial first-male paternity advantage (Gwynne, 1988b; Simmons and Achmann, 2000; Simmons et al., 2007) and variable spermatophylax sizes, perhaps as a result of variability in female availability, re-mating interval (Simmons, 1995b), and sexual status (Simmons et al., 1993). At times, therefore, gift size approximates the size necessary for complete sperm transfer.

In order to better understand the relationship between nuptial gift size and sperm transfer pattern and the selective pressure that most influences its variation, there is perhaps no better model than the bush-cricket genus *Poecilimon* (Tettigoniidae) (McCartney et al. 2008, 2012). This genus contains species with a large diversity in mating behaviours. Comparisons among species within genera can be particularly useful as characters shared by congeners are often held constant and thus control to a large degree for similarities that may be caused by relatedness (Harvey, 1991; Harvey and Pagel, 1991). With around 140 described *Poecilimon* species (Eades and Otte, 2008), the variation in nuptial gift size is unmatched among Orthoptera and approaches the magnitude of family-wide variation (McCartney et al., 2008), with spermatophore size varying from 6.1% (*Poecilimon laevissimus*) to 37% (*P. thessalicus*) of the relative body mass of the male (McCartney et al., 2008). This clearly represents large variation in male reproductive investment.

Few bush-cricket studies have investigated nuptial gift function from a sperm transfer perspective and, of these most have used laboratory-reared individuals despite concerns about the validity of this approach (see McCartney et al., 2008 for discussion). Even fewer studies still, have considered sperm transfer patterns within field populations (eg. Heller & von Helversen, 1991; Reinhold, 1994; Vahed and Gilbert, 1996). Furthermore, interpretation of data has been complicated by the diversity of taxa involved; variations in sperm transfer may ultimately be linked to taxon differences and not nuptial gift size *per se* (for discussion see Gwynne, 1995; Vahed and Gilbert, 1996; McCartney, 2010).

Our aim here was to better understand the premise that nuptial gift size relates to function. First, in order to assess the match between nuptial gift consumption time and optimum sperm transfer time across closely related species with marked variation in nuptial size, we combined published sperm transfer and nuptial gift consumption time data from two field-observed *Poecilimon* taxa that produce medium and large gifts (Reinhold and Heller, 1993, Heller & Reinhold, 1994), with sperm transfer and gift consumption data from three novel field-observed *Poecilimon* species; two with small gifts and one with very large gifts.

A close match between gift consumption and sperm transfer would be consistent with the sperm protection hypothesis, whereas if complete sperm transfer occurs long before spermatophylax gift consumption is completed, we have grounds to infer a paternal investment function. Secondly, we controlled for body mass and relatedness, and compared spermatophore size between species to the proportion of sperm that has transferred into the female by the time she has consumed the spermatophore.

A significant relationship would indicate that males of *Poecilimon* taxa that produce larger spermatophores have increased confidence of sperm transfer, and thus paternal assurance, compared to taxa producing smaller spermatophores.

Materials and Methods

Species and Sites

Poecilimon is a genus of bush-crickets (Phaneropterinae, tribe Barbistini) (Orthoptera: Ensifera: Tettigoniidae), with about 65 European species that are mostly situated in the east Mediterranean (Heller, 2004). Three species, *Poecilimon laevissimus* (Fischer, 1853), *P. erimanthos* Willemse and Heller, 1992, and *P. thessalicus* Brunner von Wattenwyl, 1891, were chosen to represent the genus in this study, as a previous study found that these species had some of the largest differences in relative spermatophore size and sperm number within the genus (McCartney et al., 2008). The spermatophore sizes of *P. laevissimus* and *P. thessalicus* represent the upper and lower limits, with *P. erimanthos* producing a small to medium-sized spermatophore of 7.2% relative mass (McCartney et al., 2008). Sperm number from single matings range between 90,000 and 140,000 - 210,000 for *P. laevissimus* and *P. erimanthos* respectively, and up to about 14,500,000 in *P. thessalicus* (McCartney et al., 2008). Data for two further species, *P. v. minor* and *P. v. veluchianus* were obtained from the literature because these species represent medium to large-size spermatophores and sperm numbers respectively (Reinhold and Heller, 1993, Heller & Reinhold, 1994). All species examined here are nocturnal except *P. erimanthos* which is diurnal and mates during the day.

Any important differences between the methods used on the novel species presented here, *P. laevissimus*, *P. erimanthos* and *P. thessalicus*, and previously published species, *P. v. veluchianus and P. v. minor*, are outlined below. However, see Reinhold and Heller (1993) and Heller and Reinhold (1994) for detailed methods on *P. v. veluchianus* and *P. v. minor*.

Fieldwork on all novel species was carried out during the summers of 1990, 1997 and 1998 on the Peloponnese Peninsula and mainland Greece. *Poecilimon erimanthos* and *P. laevissimus* were observed at Erimanthos Valley (east of the village of Kumani, N. Elia, 37°46'N, 21°47'E.), and *P. thessalicus* at a site inland from Katerini (north-west of the village of Elatochori, 40°19'N, 22°15'E). Both sites were semi-pastoral with forest margins, and population borders were demarcated by roads, forests or cliffs.

Spermatophore Consumption Time, Male Body Mass and Spermatophore Mass

All measurements on spermatophore consumption of novel species were taken from field observations of marked (*P. erimanthos*) or non marked animals (*P. laevissimus* and *P. thessalicus*) throughout their mating season. Captured animals were paired in containers or hanging mesh cages in the field. Male and female *P. laevissimus* were captured as sub-adults and allowed to mature for around seven days before pairing to allow for full development of the accessory glands (males) and full receptivity (Heller and Helversen, 1991; see Reinhold & Heller, 1993; McCartney et al., 2008 for discussion on cage and laboratory effects in *Poecilimon*). To minimize disturbance of females, observations of spermatophore consumption progress were made at intervals rather than continuously. *Poecilimon laevissimus* and *P. thessalicus* were only used in observations after witnessing the onset of spermatophore consumption, whereas we estimated onset for *P. erimanthos* as half of the interval between the first observation of a female without a spermatophore, and again with a spermatophore (females observed about every hour). Spermatophore consumption times of all species were also estimated as half of the interval between the observation of the female last seen with a spermatophore, and subsequently without a spermatophore.

Spermatophore consumption time and male body mass were measured in the 1997 and 1998 breeding seasons and pooled for *P. thessalicus* (data did not differ significantly between years; spermatophore consumption time, $t_{14} = -0.561$; $p = 0.584$; male body mass $t_{66} = -1.501$; $p = 0.138$). Spermatophore masses for *P. thessalicus* are reported from 1998. Measurements of spermatophore consumption time, male body mass and spermatophore mass for *P. laevissimus* are reported from 1997. Spermatophore consumption times were recorded for *P. erimanthos* in 1990 and the male body mass and spermatophore mass were recorded in 1997.

Sperm Transfer

Poecilimon thessalicus and *P. laevissimus* were observed in 1998 whereas *P. erimanthos* was observed in 1997 and 1998. *Poecilimon erimanthos* (in 1997) and *P. thessalicus* (in 1998) were observed at the locations where they were collected. In 1998, we collected approximately 50 sub-adult *P.*

laevissimus and *P. erimanthos* east of the village of Kumani, N. Elia and took them to Central Greece, where we made further caged observations.

All bush-crickets taken from the field were sub-adults and were stored separately by sex and species, then allowed to mature for at least seven days. We allowed mating of 20 to 30 virgin pairs of each species. Mated females were allocated randomly to predetermined spermatophore attachment times that were set at intervals relative to the spermatophore consumption time in order to determine the rate of sperm transfer. For each species, the duration of the first sperm transfer trial was set to equal the average spermatophore consumption time for that species (see Table 1). All mated females, except some *P. erimanthos* in 1997, were assigned randomly to a pre-determined transfer time for examination. For *P. laevissimus* and *P. thessalicus* we tested sperm transfer times at appropriately equal periods either side of the average spermatophore consumption time, and repeated this until we had adequately covered the full period from no transfer until (near) full transfer (*P thessalicus* = 120, 240, 480, 780, 1020, 1260 min intervals, *P. laevissimus* = 60, 120, 180, 240 min. intervals). The spermatophores of *P. erimanthos* in 1997 were removed at various intervals between 30-80 min., with two distinct modes of 35 min and 75 min This meant the mean number of sperm that had transfer in six observations between 30-35 min., and six observations between 45-80 min. were pooled into two groups at 35 min. and 80 min. and the mean sperm transfer value was used for each. The spermatophores of *Poecilimon erimanthos* in 1998 were removed at 1 min., 120 min., and 240 min and combined with the data of 1997 (35 min. and 80 min.).

Immediately after mating, each female was placed head-first into a large scintillation tube to prevent her from bending to remove the ampulla. We then stored the females in a cool, shaded area and males were returned to cages. After the assigned period, each female was removed from her vial and the spermatophore removed by grasping the ampulla at its base with dissecting forceps and pulling it carefully from her genital pore and the spermatheca was excised. The female was killed and the spermatheca and the ampulla were stored in separate vials with a known volume of water for sperm counting (1-5 ml depending on the structure's size). If sperm ampullae became semi-detached or sperm had drained outside the female these data were not used in the analysis.

Each ampulla and spermatheca was macerated with a scalpel and mixed by passing it repeatedly through a syringe until the sperm had been suspended in the water and the sample homogenised. A sub-sample was placed on a haemocytometer slide (Swift: Neubauer improved). Sperm from a minimum volume of 50 μl (or up to 200 μl) were counted and multiplied by the appropriate dilution factor to give the total number of sperm per spermatheca. Five sub-samples were taken and the solution was remixed before each new sub-sample was taken. From total sperm (ampulla and spermatheca) we derived the percentage of sperm within each mating transferred from the spermatophore into the spermatheca.

ANALYSIS

Sperm Transfer and Spermatophore Consumption

In order to compare the match/mismatch of complete sperm transfer and spermatophore consumption of all novel species, average spermatophore consumption times of all species were overlaid on a time-course chart of sperm transfer. In an attempt to compare the sperm transfer profiles of the three species presented here we spent considerable effort fitting regression models to sperm transfer patterns, and were not convinced that they could either reliably resolve the shape of sperm transfer curves, or validly explain the behaviour of sperm transferring into the female. Ultimately, no model we used could clarify the sperm transfer relationship between different species (see discussion). However, in all species examined, the modal sperm transfer time was apparent as the time when the largest change in sperm number was observed between observation intervals; in P. thessalicus this was followed by a clear plateau in the number of sperm transferred. Standard error is given in all cases.

In each species there were mating attempts resulting in no sperm transferring. These data were not included in analyses but are discussed further. Data were analysed using SAS 9.1. The analyses of *P. v veluchianus* and *P. v. minor* were as recorded in Reinhold and Heller (1993) and Heller and Reinhold (1994).

Relative Spermatophore Mass and Proportion of Sperm Transferred

Regression analyses on relative spermatophore mass against the proportion of sperm that had transferred into the female were first performed across taxa. All proportion data were arcsine (square root) transformed and tested for normality. In conjunction with this regression analysis, corresponding regression analyses were also performed on transformed proportion data with phylogenetic independent contrasts in order to control for relatedness (Felsenstein, 1985). While this method is typically preferred over standard linear regression analyses across species, sample sizes are reduced further using contrasts (n-1) and so have less power.

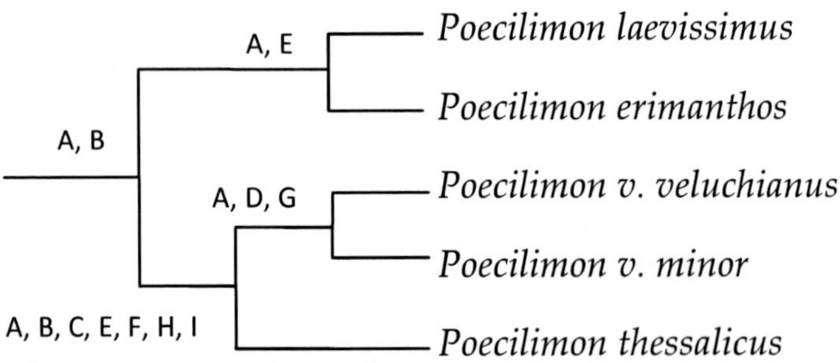

Figure 1. Cladogram representing the phylogenetic relationships between the five *Poecilimon* taxa used in this study. Letters at nodes indicate that subsequent daughter branches are based on information derived from the literature. References cited: A: Ulrich et al. 2010, B Heller 1984, C. Warchalowska-Sliwa et al. 2000, D. based on species geographic location, E. Willemse & Heller 1992, F. Heller 2006, G. Heller & Reinhold 1992, H. Heller 1990, I. Lehmann 1998.

Phylogenetic Independent Comparisons

A cladogram of the species used in this study was constructed using the literature on the phylogeny for these *Poecilimon* taxa (see Figure 1 for references) and the computer package PDAP (Maddison & Maddison, 2006) (Figure 1). The proportion of sperm and relative spermatophylax data were added to the tree in order to calculate phylogenetically-independent contrasts. In all cases, branch lengths were set to 1. The contrasts were then standardised by dividing them by the variance (square root of the sum of the branch length, Felsenstein (1985)). Generalised linear models were then used to regress the standardised independent contrasts of relative spermatophore size against the standardised independent contrasts of the proportion of transferred sperm response variable. All inferential regressions involving phylogenetically-independent contrasts were forced through the origin (Garland et al. 1992).

RESULTS

Spermatophore Consumption and Sperm Transfer

The spermatophores (spermatophylax and ampulla) were consumed in 101 ± 10.7 min (range 30-165 min., n = 14) for *P. laevissimus* (Table 1), a period too short to allow more than a small portion of the sperm to transfer into the female (Figure 2). Only about 15% of available sperm had transferred during any of the observations made prior to the last observations at 240 min (2.4 times longer than the mean spermatophore consumption period). The four observations at this longest interval revealed that a large amount of sperm still remaining in the ampulla and therefore the spermatophylax appears to be much smaller than is necessary to ensure complete sperm transfer.

Spermatophores in *P. erimanthos* (1990) were consumed in 84 ± 3.5 min (range 55-130 min, n=39, Table 1). This corresponds with a peak in sperm transfer, and more than 50% of sperm had transferred to the spermatheca by this time (Figure 2). After the time usually required for spermatophore consumption, sperm transfer seemed to slow down reaching about 75% of the total transfer after 240 min. Thirty five minutes after mating no sperm had been transferred (n=9) yet after this time all except one female (that was discarded because she was found with no sperm after 4 hours) contained over 50% (n=18) of the available sperm. So a fast transfer process occurs in this species and takes between 35 and 60 min, and is complete just before the spermatophore is normally consumed, indicating that the spermatophore may be of about the correct size for optimum sperm transfer (about 60% of total sperm). Pooled data for *P. thessalicus* from both years gave a spermatophore consumption time of 15.7 h (943 ± 47.6 min, n=16) (Table 1). The sperm transfer pattern of *P. thessalicus* differs from that in the two previous species, in that peak transfer occurred between 13-25% of mean spermatophore consumption time (240 min., Figure 2), and 93% of sperm had transferred by the end of spermatophore consumption. There was a clear plateau in sperm transfer in *P. thessalicus* at around 90-95% of total available sperm and therefore females were inseminated nearly four times more quickly than required for spermatophylax consumption. Even the fastest spermatophore consumption, of about 710 min, would have allowed around 93% of the sperm to transfer by the time one third of the spermatophore was consumed. Five out of 26 matings (19.2%) did not release any sperm into the female after transfer onset (one female at each of 240, 780, and 1260 min and two females at 480 min) and, since all other pairings resulted in close to, or above, 90% sperm transfer, these were not included in calculations of the means or standard errors. Interestingly, the average number of sperm in the ampullae that failed to transfer any sperm was only 8.3 million (n=5, S.E.=2.9 million, range = 2.3-17.8 million), significantly fewer than the 22.6 million (n=22, S.E.=21 million, range = 0.05-37.3) in spermatophores that did transfer (Mann-Whitney rank analysis U=27, $P < 0.007$).

Table 1. Male body mass, spermatophore consumption time (min) and absolute and relative spermatophore mass in three species of *Poecilimon* studied here (mean ± S.E. (range: n); upper three rows) and two sub species taken from Reinhold & Heller (1993), (lower two rows)

Species	Spermatophore consumption time (min) (range: n)	Spermatophore mass (mg)	Male body mass (mg)	Relative spermatophore size
P. erimanthos	84 ± 3.5 (55-135: 39)	47 ± 3 (n=11)	640 ± 4 (n=25)	7.2% (n=11)*
P. laevissimus	101 ± 10.7 (30-165:14)	47 ± 6 (n=9)	781 ± 13 (n=50)	6.1% (n=9)*
P. thessalicus	943 ± 47.6 (710-1380: 16)	112 ± 8 (n=28)	440 ± 7 (n=68)	33% ± 2..34% (n=17)
P. veluchianus veluchianus	570	162	640	24.9%
P. v. minor	200	74	365	19.1%

*No S.E. available because relative spermatophore mass was taken from dividing the average of pooled spermatophore mass from the average of pooled male body mass.

Relative Spermatophore Mass and Proportion of Sperm Transferred

No significant relationship was found between spermatophore size and the proportion of sperm that had transferred into the female by the spermatophore consumption time ($F_{1,4}$ = 7.69, p = 0.069, r^2 = 0.72). While this was not strengthened while controlling for relatedness ($F_{1,3}$ = 3.06, p = 0.179), a strong relationship is apparent (Figure 3). An increase in sample size is likely to produce a significant effect; males of larger spermatophore-producing *Poecilimon* taxa are likely to transfer a greater proportion of sperm than species producing smaller spermatophores.

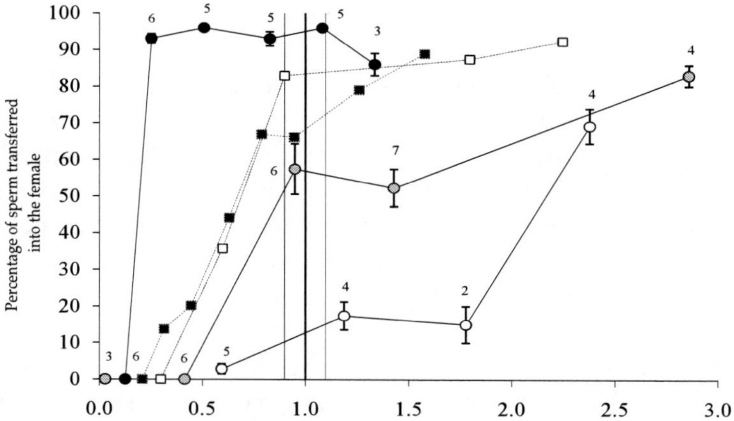

Figure 2. Percentage of sperm transferred after copulation from the male ampulla to the female spermatheca (± S.E.) plotted relative to the mean spermatophore consumption time, (numbers above points = n). Novel species are represented by unbroken lines: *P. laevissimus* (open circles), *P. erimanthos* (grey circles), and *P. thessalicus* (black circles). Broken lines represent *P. v. veluchianus* (closed squares) and *P. v. minor* (open squares) calculated from the published data (S.E and n not presented; for details see Heller and Reinhold (1994)). Dashed vertical lines show one SD in consumption time for *P. thessalicus* (the species with the largest SD).

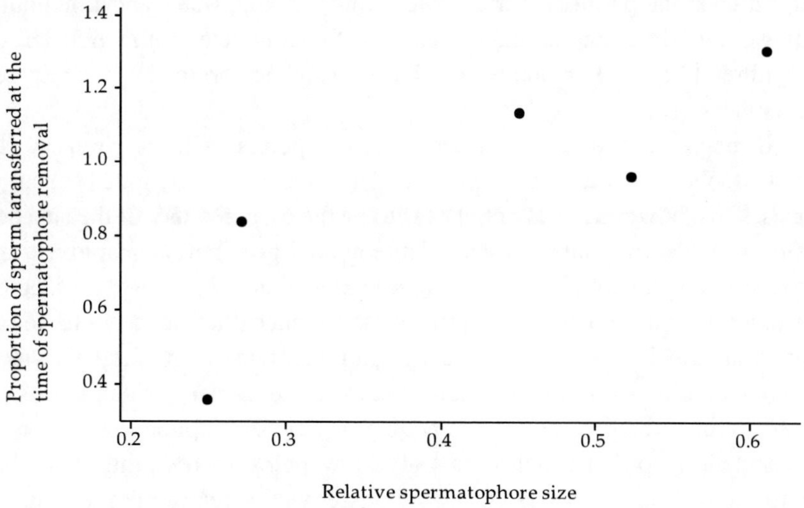

Figure 3. Spermatophore size compared to the proportion of sperm that have transferred into the female at the time of spermatophore consumption among five *Poecilimon* taxa. NB. Data are arcsine (square root) transformed.

Discussion

The percentage of sperm that had transferred into the female by the time it took her to consume and remove the spermatophore differed markedly between the five species. The match/mismatch between complete sperm transfer and spermatophore consumption time found across our species correspond to our predictions that nuptial gifts of different sizes are affected by ejaculate protection and paternal investment to different degrees. Large nuptial gifts in *Poecilimon* are apparently either of correct size or larger than necessary to allow for a full transfer of sperm, whereas small nuptial gifts seem to be less than capable of protecting the ejaculate and allowing the complete complement of sperm to be transferred. Sexual selection for larger spermatophores in *Poecilimon* is likely to increase male confidence in sperm transfer (McCartney et al., 2008, 2010) and correspond to greater level of courtship related female mating investment (McCartney et al., 2012).

Poecilimon laevissimus and *P. erimanthos* have small spermatophylaces which seem to be either smaller than necessary for sperm transfer, or have consumption times marginally correspondent with the time it takes for sperm to transfer into the female; thus they are likely to function primarily as ejaculate protection. *Poecilimon thessalicus*, on the other hand, has one of the largest spermatophores reported (McCartney et al., 2008), and a nuptial gift almost four times larger than necessary for complete sperm transfer. It is therefore likely to function as both ejaculate protection and paternal investment.

It may be assumed that *Poecilimon* species with a similarly large spermatophore size as *P. thessalicus* will also have similarly long consumption times. This, however, does not appear to be the case in either of the subspecies of *P. veluchianus* which also have large nuptial gifts but comparatively quick consumption times (Table 1, Figure 2). While *P. thessalicus* and *P. veluchianus* indeed have larger gifts, the difference does not seem to lie in the speed at which the sperm transfers into the female, but rather with the extended period over which female *P. thessalicus* consume nuptial gifts. This point is important when understanding a key assumption of the paternal investment hypothesis; males must invest in their own offspring. Lengthened consumption time increases ejaculate transfer which delays the speed at which a female re-mates and increases the time in which nutrients of the donating male's gift can be incorporated in his offspring. Reasons for an extended feeding time in *P. thessalicus* are thus far unknown, however the study population was at relatively high altitude (ca. 1,100 m a.s.l.), with night time temperature often at around 10-15°C, compared to *P. v. veluchianus* and *P. v. minor* (330 m a.s.l.; Reinhold and Heller, 1993) and *P. erimanthos* and *P. laevissimus* (around 600 m a.s.l.) where night temperatures are typically 20°C or above (unpubl. data). The metabolism of *P. thessalicus* at these temperatures is likely to be lower than that of the other species, resulting in consumption duration and digestion times of nuptials gift being significantly slower. However, temperature differences are unlikely to have affected our results because spermatophores are costly to produce and are evolutionary labile (McCartney et al., 2008); males would be expected to allocate fewer resources to gift production – to a size more appropriate to ejaculate protection – if there were no fitness benefits to having a proportionately large spermatophylax gift. A further explanation for the slow gift consumption time of *P. thessalicus* may be related to possible bitter substances in the spermatophylax of *P. thessalicus*. These may affect the speed at which females are able to consume the nuptial gift (as suggested by Heller et al., 1998) but further work is needed in order to verify the substances, their palatability, and the effect they have on females.

While there is a match between nuptial gift consumption and sperm transfer times in *P. v. veluchianus* and *P. veluchianus minor* these species have nuptial gift sizes that further correspond to our predictions that larger gifts are influenced by paternal investment. *Poecilimon v. veluchianus* has a substantial last male mating advantage (Achmann et al., 1992) and nutrients from the nuptial gift of the donating male are likely invested into his own offspring (McCartney 2010). Evidence of paternal investment in this species comes from a correlation of nuptial gift size on the dry mass of the donating males' offspring, and a greater lifespan of starved offspring (immediately after eclosion) fathered by males with large spermatophores (Reinhold, 1999).

Typically it takes 3-4 days for the nutrients in nuptial gifts to be incorporated in egg batches in the female (Bowen et al., 1984; Gwynne and Brown, 1994; Wedell, 1993b; Simmons and Gwynne, 1993; Reinhold, 1999, although see Voigt et al., 2006, 2008). While sperm precedence patterns have only been analysed in two *Poecilimon* species (*P. veluchianus*; Achmann et al., 1992 and *Poecilimon hoelzeli*; Achmann, 1996), both show a last male mating advantage. The combination of these factors in both *P. erimanthos* and *P. laevissimus* is, however, likely to exclude the possibility of paternal investment; both species remate, on average, every 1-2 days and lay eggs every two days (McCartney 2010). It is therefore unlikely that there is sufficient time for gift nutrients to be incorporated into the donating male's offspring. In contrast, field observations from *P. thessalicus* suggest that females may have extended inter-mating refractory periods of about 7-8 days (and up to 19 days) and lay eggs every 1-2 days (McCartney, 2010), so males are likely to have their nutrients incorporated into the majority of eggs before females remate.

Transfer of a full ejaculate is necessary to ensure optimum fertilisation for males especially in polyandrous species (Smith, 1984). It is difficult therefore to understand why, in the two species that produce smaller spermatophores, males do not protect their ejaculate with larger nuptial gifts. In *P. erimanthos* and *P. laevissimus* females consumed 48% and 87% respectively of the sperm that males produced. While *P. laevissimus* seemed to remove and eat the spermatophore nearly eight times faster than expected for maximum sperm transfer, sperm from *P. erimanthos* transferred into the female at a rate that arguably approximated spermatophore consumption time, but still resulted in a waste of a large portion of sperm. Similarly, 9% of spermatophores are estimated to be prematurely consumed in *P. v. veluchianus* (Reinhold and Heller, 1993). It is likely that there is conflict between the sexes over optimum sperm number and resulting spermatophore attachment duration. Premature removal of the ampulla may constitute a form of post-copulatory female discrimination (Sakaluk and Eggert, 1996), but it is unlikely that such a high number of matings observed here resulted in removal discrimination. Sperm loading, the adjustment of copulation duration and ejaculate size according to the risk of sperm competition (Parker et al., 1990), has been observed in some other insects species (see for example Dickenson, 1986; Garcia-Gonzalez and Gomendio, 2004) including bush-crickets *Uromenus rugosicollis* (Vahed, 1997), and may well be a feature in some *Poecilimon*. Males may produce an optimum number of sperm ideal for sperm competition but in *P. laevissimus*, females may "have the edge" over this conflict by being able to consistently consume and remove the nuptial gift and sperm ampulla before the sperm is fully transferred (reviewed in Vahed, 2007b; Gwynne, 2008). This assertion of a conflict between the sexes is further corroborated by evidence in *P. laevissimus* where the pairs struggle for some time as the females appear to try and escape the clasp of the male's cerci, and may additionally represent a form of female discrimination that leads to 'fit' males transferring more sperm overall (Eberhard, 1996).

In a different form, large quantities of sperm and spermatophore material are also wasted in *P. thessalicus*. We found that a large proportion of males did not transfer any sperm (18.5%; n=27). Spermatophores are expensive to produce (Dewsbury, 1982; Drummond, 1984; Simmons, 1990, 1995a; Heller and von Helversen, 1991; Vahed, 2007b; Lehmann, 2012), so those that fail to initiate represent a considerable waste in time and energy to *P. thessalicus* males. It may be that constriction of the females in scintillation tubes affected the onset of sperm transfer in *P. thessalicus*, although this is unlikely as onset was not affected in *P. laevissimus, P. erimanthos* and a previously studied species, *P. hoelzeli* (R. Achmann *pers. comm.*). Importantly, total sperm numbers in ampullae that did not transfer were much less (by 63%) than the total number of sperm in ampulla that did transfer. While these data suggest that the mechanical initiation of sperm transfer may be dependent on the internal pressure or volume of sperm or ejaculate, mechanisms behind the sperm transfer process are poorly understood in bush-crickets. Future studies would benefit by further assessment of sperm transfer initiation during this critical onset period (Achmann et al., 1992; Reinhold and Heller, 1993; Simmons and Achmann, 2000; Simmons, 2001).

Ultimately, no model we used for analysis could clarify the sperm transfer relationship between different species. Vahed (1994), however, previously fitted models using data from Gwynne et al. (1984) and Gwynne (1986) and showed that there was no difference between the sperm transfer curves for *Leptophyes punctatissima* and *Requena verticalis*, two species with varied sperm transfer profiles. Vahed (1994) suggested that the variation found within the sperm transfer among individuals of each species may be too large to easily detect a difference among species, although ultimately concluding that the function of the spermatophylax in *R. verticalis* is likely the same as that for *L. punctatissima*; to protect the ejaculate. As a comparison, we adopted the model used by Vahed (1994) and similarly found no difference between the sperm transfer curves of the two most different *Poecilimon* species (i.e. *P. thessalicus* and *P. laevissimus*, S=0.261, P=0.61). We therefore suggest that the variation found within the sperm transfer among individuals of each species is too large to detect a difference between species and that the curves are unlikely to be considered the same.

It is important to keep in mind that the function of the nuptial gift is influenced by substances in the ampulla, other than sperm, that are transferred during mating (McCartney et al. 2008; McCartney, et al. 2010). Some of these substances are known to influence female intermating refractory period (Heller & Helversen, 1991; Heller & Reinhold, 1994; Lehmann and Lehmann, 2000b; Vahed, 2007b), the timing of oviposition (Arnqvist and Rowe, 2005; Vahed, 2007b), and the share of eggs that are laid with the donating male's nutritional investment (Simmons 1990; Vahed, 2003). Indeed, the positive relationship we found between spermatophore size and the proportion of sperm transferred may tie closely to the total volume of ejaculate substances transferred. If these substances affect fertilisation success or the incorporation of nutrients into offspring, the size or function of the nuptial gift may instead vary in accordance with these and be an important factor governing gift size (Vahed, 2003, 2007a; McCartney et al., 2008).

It is unlikely that male *P. erimanthos* or *P. laevissimus* make significant paternal investments in their offspring in terms of nutrients. While paternal investment has been directly observed in *P. v. veluchianus* (Reinhold 1999), the disparity in time between complete sperm transfer and spermatophore consumption in *P. thessalicus* is also best explained by paternal investment. Larger spermatophores apparently increase male confidence in sperm transfer – and perhaps total ejaculate transfer – and are likely to ensure a greater level of paternal assurance. Furthermore, a recent study has shown that females of *Poecilimon* species that have males that invest more in spermatophore production will compete for access to males, invest relatively more in mating effort, and take greater risks in finding mates than species with smaller spermatophores (McCartney et al. 2012).

In terms of ejaculate protection and paternal investment, we present evidence that both sexual selection and natural selection influence spermatophore size within the single bush-cricket genus *Poecilimon*. However, irrespective of function, it is clear that sperm is wasted in all species presented here, and a better understanding is needed of the cost of sperm production as well as the mechanisms which affect sperm transfer if we are to fully understand the relationship between nuptial gift size, paternal investment, and ejaculate protection. Future studies would do well to assess how other substances in the ejaculate may control female re-mating, ova production and oviposition rate, and how the transference of these substances relate to gift consumption time.

ACKNOWLEDGMENTS

We thank K. Teltscher, L. Penny, M. Volleth and K. Witt for help in the field, K. Reinhardt, K. Reinhold and W. Edrich for discussion and, K. Vahed and M. Rossiter for helpful comments on earlier versions of the manuscript. Our research was supported by Massey University and D.F.G. (Deutsche Forschungsgemeinschaft) scholarships.

REFERENCES

Achmann R. 1996. Individueller Paarungs- und Fortpflanzungserfolg bei der Laubheuschrecke *Poecilimon hoelzeli* : Ein molekulargenetischer Ansatz. PhD thesis. Friedrich-Alexander-Universität.

Achmann R, Heller K-G, Epplen JT. 1992. Last-male sperm precedence in the bushcricket *Poecilimon veluchianus* (Orthoptera, Tettigonioidea) demonstrated by DNA fingerprinting. *Molecular Ecology* 1: 47-54.

Arnqvist G, Rowe C. 2005. *Sexual conflict (Monographs in Behavior and Ecology)*. Princeton, NJ: Princeton University Press.

Boldyrev BT. 1915. Contributions à l'étude de la structure des spermatophores et des particularitès de la copulation chez Locustodea et Gryllodea. *Horae Societatis Entomologicae Rossicae* 6: 1-245.

Bowen BJ, Codd CG, Gwynne DT. 1984. The katydid spermatophore (Orthoptera: Tettigoniidae): male investment and its fate in the mated female. *Australian Journal of Zoology* 32: 23-31.

Dewsbury DA. 1982. Ejaculate cost and male choice. *American Naturalist.* 119: 601-610.

Dickinson JL. 1986. Prolonged mating in the milkweed leaf beetle *Labidomera clivicollis clivicollis* (Coleoptera: Chrysomelidae): a test of the "sperm-loading" hypothesis. *Behavioral Ecology and Sociobiology* 18: 331-338.

Drummond BA ed. 1984. *Multiple mating and sperm competition in the Lepidoptera*. New York: Academic Press.

Eades DC, Otte. D. 2008. Orthoptera Species File Online. Version 2.0/3.4. http://Orthoptera.SpeciesFile.org.

Eberhard WG. 1996. *Female control: sexual selection by cryptic female choice*. Princeton, NJ: Princeton University Press.

Felsenstein J. 1985. Phylogenies and the comparative method. *American Naturalist* 125: 1-15.

Garcia-Gonzalez F, Gomendio M. 2004. Adjustment of copula duration and ejaculate size according to the risk of sperm competition in the golden egg bug (Phyllomorpha laciniata). *Behavioral Ecology* 15: 23-30.

Gerhardt U. 1913 Copulation und Spermatophoren von Grylliden und Locustiden I. *Zoologische Jahrbucher (Syst.)* 35: 415-532.

Gwynne DT. 1984. Sexual selection and sexual differences in mormon crickets (Orthoptera: Tettigoniidae, *Anabrus simplex*). *Evolution* 38: 1011-1022.

Gwynne DT. 1986. Courtship feeding in katydids (Orthoptera, Tettigoniidae) - investment in offspring or in obtaining fertilizations. *American Naturalist* 128: 342-352.

Gwynne DT. 1988a. Courtship feeding and the fitness of female katydids (Orthoptera: Tettigoniidae). *Evolution* 42: 545-555.

Gwynne DT. 1988b. Courtship feeding in katydids benefits the mating males Offspring. *Behavioral Ecology and Sociobiology* 23: 373-377.

Gwynne DT. 1990. The katydid spermatophore: evolution of a parental investment. In: Bailey WD and Rentz DC, eds. *The tettigoniidae: biology, systematics and evolution.* Bathurst: Crawford house press. 27-240.

Gwynne DT. 1997. The evolution of edible 'sperm sacs' and other forms of courtship feeding in crickets, katydids and their kin (Orthoptera: Ensifera). In: Choe JC and Crespi BJ, eds. *Evolution of mating systems in insects and arachnids*: University of Cambridge. 110-129.

Gwynne DT. 2001. *Katydids and bush-crickets: reproductive behavior and evolution of the tettigoniidae.* Cornell University Press: Ithaca.

Gwynne DT. 2008. Sexual conflict over nuptial gifts in insects. *Annual Review of Entomology.* 53: 83-101.

Gwynne DT, Bowen BJ, Codd CG. 1984. The function of the katydid spermatophore and its role in fecundity and insemination ((Orthoptera: Tettigoniidae). *Australian Journal of Zoology* 32: 15-22.

Gwynne DT, Brown WD. 1994. Mate feeding, offspring investment, and sexual differences in katydids (Orthoptera: Tettigoniidae). *Behavioral Ecology* 5: 267-272.

Harvey PH. 1991. Comparing uncertain relationships: The Swedes in revolt. *Trends in Ecology & Evolution* 6: 38 - 39.

Harvey PH, Pagel MD. 1991. *The comparative method in evolutionary biology*. Oxford University Press: Oxford.

Heller K-G. 1984. To the bioacoustics and phylogeny of the genus *Poecilimon* (Orthoptera, Tettigoniidae, Phaneropterinae). *Zoologische Jahrbücher. Systemologie* 111: 69-117.

Heller K-G. 1990. Evolution of song pattern in east Mediterranean Phaneropteridae: constraints by the communication system. In: Bailey WJ and Rentz DCF, eds. *The Tettigoniidae: Biology, Systematics and Evolution*. Bathurst, Australia.: Crawford House. 130-151.

Heller K-G. 1992. Risk shift between males and females in the pair-forming behavior of bush-crickets. *Naturwissenschaften* 79: 89-91.

Heller K-G. 2004. Fauna europaea: insecta: phaneropteridae. fauna europaea version 1.1 *Fauna Europaea*. 1.1 ed. http://www.faunaeur.org.

Heller K-G, Faltin S, Fleischmann P, von Helversen O. 1998. The chemical composition of the spermatophore in some species of phaneropterid bush crickets (Orthoptera: Tettigonioidea). *Journal of Insect Physiology* 44: 1001-1008.

Heller K-G, Korsunovskaya O, Sevgili H, Zhantiev RD. 2006. Bioacoustics and systematics of the *Poecilimon heroicus* group (Orthoptera: Phaneropteridae: Barbitistinae). *European Journal of Entomology.* 103: 853-865.

Heller K-G, Reinhold K. 1994. Mating effort function of the spermatophore in the bushcricket *Poecilimon veluchianus* (Ortoptera, Phaneropteridae): support from a comparison of the mating behaviour of two subspecies. *Biological Journal of the Linnean Society* 53: 153-163.

Heller K-G, von Helversen D. 1991. Operational sex-ratio and individual mating Frequencies in two bush-cricket species (Orthoptera, Tettigonioidea, *Poecilimon*). *Ethology* 89: 211-228.

Lehmann AW. 1998. Artbildung, akustische Kommunikation und sexuelle Selektion bei griechischen Laubheuschrecken der *Poecilimon propinquus* Gruppe (Orthoptera: Phaneropteridae).

Lehmann, GUC. Lehmann, A.W. 2000. Female bushcrickets mated with parasitized males show rapid remating and reduced fecundity (Orthoptera: Phaneropteridae: *Poecilimon mariannae*). *Naturwissenschaften* 87: 404-407.

Lehmann, GUC. 2012. Weighing costs and benefits of mating in bushcrickets (Insecta: Orthoptera: Tettigoniidae) with an emphasis on nuptial gifts, protandry and mate density. *Frontiers in Zoology* 9: 19.

McCartney J, Heller K-G, Potter MA, Robertson AW, Telscher K, Lehmann G, Lehmann A, Helversen D v, Reinhold K, Achmann R. 2008. Understanding nuptial gift size in bush-crickets: an analysis of the genus *Poecilimon* (Tettigoniidae; Orthoptera). *Journal of Orthoptera Research* 17: 231-242.

McCartney J, Lehmann AW, Lehmann GUC. 2010. Lifetime spermatophore investment in natural populations of two closely related bush-cricket species (Orthoptera: Tettigoniidae: *Poecilimon*). *Behaviour* 147: 285-298.

McCartney J. 2010. Spermatophore size variation in the bush-cricket genus PhD, Massey University, New Zealand: pp 297.

McCartney J, Kokko H., Heller K-G, Gwynne D. 2012. The evolution of sex differences in mate searching when females benefit: new theory and a comparative test. *Proceedings of the Royal Society of London (B).* 279: 1225 - 1232.

Parker GA, Simmons LW, Kirk H. 1990. Analysing sperm competition data: simple models for predicting mechanisms. *Behavioral Ecology and Sociobiology* 27: 55-65.

Quinn JS, Sakaluk SK. 1986. Prezygotic male reproductive effort in insects: why do males provide more than sperm. *Florida Entomologist* 69: 84-94.

Reinhold K. 1999. Paternal investment in *Poecilimon veluchianus* bushcrickets: beneficial effects of nuptial feeding on offspring viability. *Behavioral Ecology and Sociobiology* 45: 293-299.

Reinhold K, Heller K-G. 1993. The ultimate function of nuptial feeding in the bushcricket *Poecilimon veluchianus* (Orthoptera: Tettigoniidae: Phaneropterinae). *Behavioral Ecology and Sociobiology* 32: 55-60.

Sakaluk SK. 1984. Male crickets feed females to ensure complete sperm transfer. *Science* 223: 609-610.

Sakaluk SK, Eggert AK. 1996. Female control of sperm transfer and intraspecific variation in sperm precedence: Antecedents to the evolution of a courtship food gift. *Evolution* 50: 694-703.

Simmons LW. 1990. Nuptial feeding in Tettigoniids: male costs and the rates of fecundity increase. *Behavioral Ecology & Sociobiology* 27: 43-47.

Simmons LW. 1995a. Courtship feeding in Katydids (Orthoptera: Tettigoniidae): Investment in offspring and in obtaining fertilizations. *The American Naturalist* 146: 307-315.

Simmons LW. 1995b. Male bushcrickets tailor spermatophores in relation to their remating intervals. *Functional Ecology* 9: 881-886.

Simmons LW. 2001. *Sperm competition and its evolutionary consequences in the Insects*. Princeton University Press: Princeton, NJ.

Simmons LW, Parker PG. 1989. Nuptial feeding in insects: Mating effort versus paternal investment. *Ethology* 81: 332-343.

Simmons LW, Gwynne DT. 1991. The refractory period of female katydids (Orthoptera: Tettigoniidae): sexual conflict over the remating interval? *Behavioral Ecology* 2: 276-282.

Simmons LW, Craig M, Llorens T, Schinzig M, Hosken D. 1993. Bushcricket spermatophores vary in accord with sperm competition and parental investment theory. *Proceedings for the Royal Society of London* 251: 183-186.

Simmons LW, Gwynne DT. 1993. Reproductive investment in bushcrickets: the allocation of male and female nutrients to offspring. *Proceedings of the Royal Society of London (B)* 252: 1-5.

Simmons LW, Beesley L, Lindhjem P, Newbound D, Norris J, Wayne A. 1999. Nuptial feeding by male bushcrickets: an indicator of male quality? *Behavioral Ecology* 10: 263-269.

Simmons LW, Achmann R. 2000. Microsatellite analysis of sperm-use patterns in the bushcricket *Requena verticalis*. *Evolution* 54: 942-952.

Simmons LW, Beveridge M, Kennington WJ. 2007. Polyandry in the wild: temporal changes in female mating frequency and sperm competition intensity in natural populations of the tettigoniid *Requena verticalis*. *Molecular Ecology* 16: 4613-4623.

Smith RJ. 1984. Sperm utilisation strategies of females. In: Smith RL, ed. *Sperm Competition and the Evolution of Animal Mating Systems*: Academic Press Press.

Thornhill R. 1976. Sexual selection and paternal investment in insects. *American Naturalist* 110: 153-163.

Thornhill R, Alcock J. 1983. *The Evolution of Insect Mating Systems*. Harvard University Press: Cambridge, MA.

Trivers RL. 1972. Parental investment and Sexual Selection. In: Campbell B, ed. *Sexual Selection and the Descent of Man*. Chicago.: Aldine.

Ulrich B, Reinhold K, Niehuis O., Misof B. 2010. Secondary Structure and Phylogenetic Analysis of the Internal Transcribed Spacers 1 and 2 of Bush Crickets (Orthoptera: Tettigoniidae: Barbitistini). *Journal of Zoological Systematics and Evolutionary* Research: 48, 219-228.

Vahed K. 1994. The evolution and function of the spermatophylax in bushcrickets (Orthoptera: Tettigoniidae). PhD Thesis, University of Nottingham.

Vahed K. 1997. Copulation and spermatophores in the Ephippigerinae (Orthoptera: Tettigoniidae): prolonged copulation is associated with a smaller nuptial gift in *Uromenus rugosicollis* Serville. *Journal of Orthoptera Research* 6: 83-89.

Vahed K. 1998. The function of nuptial feeding in insects: review of empirical studies. *Biological Reviews* 73: 43-78.

Vahed K. 2003. Increases in egg production in multiply mated female bushcrickets *Leptophyes punctatissima* are not due to substances in the nuptial gift. *Ecological Entomology* 28: 124-128.

Vahed K. 2007a. Comparative evidence for a cost to males of manipulating females in bushcrickets. *Behavioral Ecology* 18: 499-506.

Vahed K. 2007b. All that glisters is not gold: sensory bias, sexual conflict and nuptial feeding in insects and spiders. *Ethology* 113: 105-127.

Vahed K, Gilbert FS. 1996. Differences across taxa in nuptial gift size correlate with differences in sperm number and ejaculate volume in bushcrickets (Orthoptera: Tettigoniidae). *Proceedings of the Royal Society of London: B* 263: 1257-1265.

Voigt CC, Kretzschmar AS, Speakman JR, Lehmann GUC. 2008. Female bushcrickets fuel their metabolism with male nuptial gifts. *Biology Letters* 4: 476-478.

Voigt CC, Lehmann GUC, Michener RH, Joachimski MM. 2006. Nuptial feeding is reflected in tissue nitrogen isotope ratios of female katydids. *Functional Ecology* 20: 656-661.

Warchalowska-Sliwa E, Heller KG, Maryanska-Nadachowska A, Lehmann A. 2000. Chromosome evolution in the genus *Poecilimon* (Orthoptera, Tettigonioidea, Phaneropteridae). *Folia Biologica-Krakow* 48: 127-136.

Wedell N. 1991. Sperm competition selects for nuptial feeding in a bushcricket. *Evolution* 45: 1975-1978.

Wedell N. 1993a. Spermatophore size in bushcrickets: comparative evidence for nuptial gifts as a sperm protection device. *Evolution* 47: 1203-1212.

Wedell N. 1993b. Mating effort or paternal investment? Incorporation rate and cost of male donations in the wartbiter. *Behavioral Ecology and Sociobiology* 32: 239-246.

Wedell N. 1994a. Variation in nuptial gift quality in bush crickets (Orthoptera: Tettigoniidae). *Behavioral Ecology* 5: 418-425.

Wedell N. 1994b. Dual function of the bushcricket spermatophore. *Proceedings of the Royal Society of London* 258: 181-185.

Wedell N, Arak A. 1989. The wartbiter spermatophore and its effect on female reproductive output (Orthoptera: Tettigioniidae, *Decticus verrucivorus*. *Behavioral Ecology and Sociobiology* 24: 117-125.

Willemse F, Heller K-G. 1992. Notes on Systematics of Greek species of *Poecilimon* Fischer, 1835 (Orthoptera: Phaneropteridae). *Tijdschrift voor Entomologie* 135: 299-315.

In: Sexual Selection
Editors: R. Geldani and M. Davin
ISBN: 978-1-62808-805-2

Chapter 3

MATE CHOICE COPYING IN BOTH SEXES OF THE GUPPY: THE ROLE OF SPERM COMPETITION RISK AND SEXUAL HARASSMENT

Claudia Zimmer*[*,1], Alexandros S. Gavalas[1], Benjamin Kunkel[1], Janina Hanisch[1], Sara Martin[1], Svenja Bischoff[1], Martin Plath[1] and David Bierbach[1,2]

[1]Department of Ecology and Evolution, University of Frankfurt, Frankfurt am Main, Germany

[2]Leibniz-Institute of Freshwater Ecology and Inland Fisheries, Berlin, Germany

ABSTRACT

Mate choice copying was mostly described as a strategy employed by females to assess the quality of potential mates, but also males can copy other males' mate choice. In both cases, focal individuals show an increased propensity to copulate with a potential mating partner they could observe interact sexually with another individual (the 'model'). Sexual interactions, however, convey additional—partly conflicting—information to the choosing individual: females may try to avoid sexually

* Author to whom correspondence should be addressed; E-Mail: cla.zim.uni@gmail.com (C.Z.).

active males due to a rougher courtship or coercive mating attempts, and males may respond to an increased sperm competition risk (SCR). How do females and males respond to different copying situations, in which the model individual either resides in the vicinity of a potential mating partner without physical contact (i.e., no harassment, and low SCR), or interacts sexually with the potential mating partner? Do individuals copy less in the latter situation? We investigated these questions in the guppy (*Poecilia reticulata*), a livebearing fish with internal fertilization, strong SCR among males, and frequent sexual harassment of females. Focal individuals could choose to associate with a large or a small stimulus fish, and mate choice tests were repeated after the previously non-preferred stimulus fish could be seen associating (low harassment and low SCR) or physically interacting (high harassment and high SCR) with a model individual. In a control treatment, no model was presented. We found both males and females to copy similarly in both copying treatments, while no response was observed in the control. This contrasts with a study reporting that Atlantic molly (*P. mexicana*) males copy less under elevated SCR. Even though we lack a compelling explanation as to why both congeners might differ, we are tempted to argue that strong(er) benefits arising from copying may have selected guppies to copy in a broader range of contexts, including situations where the choosing individual incurs harassment or SCR.

Keywords: Non-independent mate choice, Sexual selection, Public information, Social learning, Sperm competition, Mate choice copying

INTRODUCTION

Sexual selection, e.g., through mate choice, is a powerful evolutionary force, and individuals may integrate a wide range of information in their assessment of mate quality (Andersson 1994; Candolin 2003). Besides phenotypic traits like body coloration (Houde 1997; Price et al. 2008) and body size (Reznick and Endler 1982), dynamic traits like courtship displays (Darrow and Harris 2004; Rosenthal 2007) and parental care (Warner et al. 1995), also social interactions with conspecifics other than the choosing individual may serve as cues indicating mate quality (Ophir & Galef jr 2004; Ziege et al. 2012; Bierbach et al. 2013a). Especially in group-living animals that spend most of their lives surrounded by conspecifics, mating takes place not in privacy, but in a public domain (Valone and Templeton 2002; Danchin 2004; Dabelsteen 2005; Valone 2007; Druen and Dugatkin 2011). McGregor and Peake (2000), introduced the term 'communication networks' to account for the multi-directional nature of communicative interactions in animal societies. In such communication networks, by-standing individuals not involved in the sender-receiver dyad (but residing within a physical range allowing them to detect a signal), have the possibility to obtain information by observing communicatory interactions amongst other group members.

The process of extracting information from social interactions among other individuals within a communication network has been termed 'social eavesdropping' (Naguib et al. 2004; Dabelsteen 2005; Peake 2005; Valone 2007; Fitzsimmons et al. 2008). A widespread form of social eavesdropping in the context of mate choice involves the observation of aggressive male conflicts by females and subsequent choice of one of the rivals based on its performance (Ophir and Galef jr 2003; Aquiloni et al. 2008; Bierbach et al. 2013a). Another mechanism involves the observation of sexual interactions between males and females and subsequent choice of one of the involved individuals; termed 'mate choice copying' (Dugatkin 1992; Pruett-Jones 1992; Nordell and Valone 1998; Westneat et al. 2000; Bonnie and Earley 2007; Witte and Nöbel 2011).

Mate choice copying has been reported for various vertebrate taxa like birds (Höglund et al. 1990; Galef jr and White 1998; White and Galef jr 1999; Swaddle et al. 2005; Freed-Brown and White 2009) and mammals (Clutton-Brock and McComb 1993; Galef jr et al. 2008), including humans (Uller and Johansson 2003; Waynforth 2007; Yorzinski et al. 2010), but also in invertebrates (Mery et al. 2009; but see Auld et al. 2009). However, following Darwin's groundbreaking work on sexual selection (Darwin 1871) most studies assumed females to be the choosing sex due to their higher investment in oocyte production compared to males (Andersson 1994). As a consequence, mate choice copying was most extensively studied in females, especially in teleost fishes like guppies, *Poecilia reticulata* (Dugatkin 1992, 1996, 1998; Dugatkin and Godin 1992, 1993; Brooks 1999; Godin and Hair 2009), sailfin mollies, *Poecilia latipinna* (Schlupp et al. 1994; Witte and Ryan 1998, 2002; Witte and Noltemeier 2002; Witte and Ueding 2003), Hispañola limia, *Limia perugiae* (Applebaum and Cruz 2000), Japanese medaka, *Oryzias latipes* (Grant and Green 1996) and Atlantic mollies, *Poecilia mexicana* (Heubel et al. 2008).

In theory, females can benefit from mate choice copying in at least two ways (after Gibson and Höglund 1992): (1) The accuracy of their own mate assessment may be increased, which provides a good opportunity for young females to copy decisions of older—more experienced—females (Dugatkin and Godin 1993). (2) Moreover, the time spent in assessing mate quality may be reduced (see Witte and Nöbel 2011 for discussion). The importance of female mate choice copying in shaping mating preferences can be illustrated in sailfin mollies (*P. latipinna*), in which genetically-based female preferences, such as the innate preference for large male body size, are reduced or even reversed through mate choice copying (Witte and Noltemeier 2002).

In promiscuous species, sexual activity (or the number of mating partners) may serve as an indicator of male quality. For example, females of the Atlantic molly were recently shown to have an increased propensity of interacting with males they saw mating with either a female or with another male, and the authors concluded that male sexual activity, whether in a heterosexual or homosexual context, is indeed a trait females find attractive (Bierbach et al. 2013b). However, poeciliid males exhibit high frequencies of sexual behaviors (Farr 1989; Magurran and Seghers 1994; Bisazza and Marin 1995; Houde 1997; Magurran 2001; Plath et al. 2003, 2007), while females can store sperm to fertilize a succession of broods and require only few copulations (Constantz 1984, 1989; Magurran and Seghers 1994; Pilastro et al. 2007). High frequencies of male sexual behavior are known to cause physical damage to the female genital tract (Magurran 2011), and females have less time available for feeding in presence of a harassing male as females spent considerable time trying to avoid sexual harassment (Magurran and Seghers 1994; Plath et al. 2007; Köhler et al. 2011). So, associating with a male that proofed its sexual activity (or at least its motivation to mate through its proximity to another female) obviously brings about an increased risk of sexual harassment. In the current study we asked whether females of two populations of guppies copy less under an increased level of sexual harassment (e.g., through observing sexual interactions between a male and a model female vs. mere associations between both).

For some species, also male mate choice copying has been described [sailfin mollies, *P. latipinna* (Schlupp and Ryan 1997; Witte and Ryan 2002), pipefish, *Syngnathus typhle* (Widemo 2005) and three-spined stickleback, *Gasterosteus aculeatus* (Frommen et al. 2009)]. However, male mate copying in internally fertilizing species—like mammals or poeciliid fishes—remains a conundrum as males incur an increased risk of sperm competition when choosing another male's previous mate (sensu Parker 1979; e.g., Constantz 1984; Becher and Magurran 2004). Nevertheless, poeciliid females are most receptive as virgins or for few days postpartum (Rosenthal 1952; Constantz 1984; Liley 1996) and, accordingly, only a small proportion of females in a population is receptive at the same time (Houde 1997; Magurran 2005). Furthermore, males evaluate female receptivity through so-called nipping behavior, where a male touches the female's genital opening with its snout (Parzefall 1969). Alternatively, males can observe sexual interactions of other males with females, and copying other males' mate choice may allow saving considerable energetic and opportunity costs associated with mate searching and testing different females via nipping (sensu Schlupp and Ryan 1997). Additionally, in some poeciliids (including the guppy), last male sperm precedence is known (Farr 1989; Evans and Magurran 2001; Becher and Magurran 2004), which gives a competitive edge to the copying male even if the copied one has previously inseminated the female. Nevertheless, sperm competition plays a crucial role in poeciliid mating behavior (see also Evans and Magurran 2001; Dosen and Montgomerie 2004; Wong and McCarthy 2009; Evans and Pilastro 2011; Jeswiet et al. 2011), and a recent study by Bierbach et al. (2011a) found Atlantic molly (*P. mexicana*) males to copy less when sperm competition risk was high (e.g., when males observed sexual interactions between the model male and a stimulus female rather than mere associations without copulation).

While mate choice copying in general appears to be beneficial for both, males and females, risk of sexual harassment (for females) or risk of sperm competition (for males) ought to render certain copying situations less attractive. Specifically, we would expect a fine-tuning of mate choice copying behavior according to the nature of the observed interactions. In the current study, we compared male and female mate choice copying in guppies between two treatments (following the protocol of Bierbach et al. 2011a) during which focal individuals could either observe mere associations between a stimulus fish and a model (representing no sexual harassment or sperm competition) or direct sexual interactions (high risk of sexual harassment or sperm competition). Specifically, we asked (a) whether males and females show differences in their mate choice copying behavior and (b) whether the copying situation (no vs. high risk of sexual harassment or sperm competition) affects the strength of copying. In another experiment, we asked if males and females have a predilection for either of the two copying situations. We thus presented another set of males and females with video sequences showing a male or a female in association or engaging in sexual behavior and measured their preferences for both types of copying situations.

Materials and Methods

Study Species and Animal Maintenance

Test fish came from two populations of guppies (*P. reticulata*): individuals from the first population were second to third generation descendants of wild-caught fish originating from Venezuela and were imported to Germany by '*Aquarium Dietzenbach*' (coded as 'wild-type' in the following sections), while fish from the second population were regular aquarium fish (coded as 'domestic' in the later). Fish were reared in mixed-sex, single-population stock tanks (150-200 L) under a 12:12 light:dark cycle at a constant temperature of 28°C. All tanks were equipped with natural gravel and artificial plants to provide shelter. We fed the fish to satiation with commercially available flake food (TetraMin®) and food tablets (TetraTabs®) twice daily, and frozen chironomid lavae and *Artemina salina* nauplii once a week.

Experiment 1

To compare male and female mate choice copying in different copying situations (see Introduction), we largely followed the experimental protocol of Schlupp and Ryan (1997) that was recently modified and used to investigate male mate choice copying under sperm competition risk in the Atlantic molly (*P. mexicana*, Bierbach et al. 2011a). All focal fish were isolated prior to the tests in groups of < 20 individuals in 100-L aquaria for at least 24 hours. Focal and stimulus individuals stemmed from different isolation groups to prevent effects of familiarity (Bierbach et al. 2011b). A standard test tank (80 cm length × 30 cm height × 30 cm depth) was divided into five sections of equal size: The two lateral compartments were divided by transparent Plexiglas partitions and contained the stimulus fish (Fig. 1). The remaining compartment was visually divided in two lateral preference zones and a central neutral zone by marks drawn on the front of the tank. The tank was filled to a height of 15 cm with aged tap water of 27–29°C and was aerated and heated whenever no experiment was conducted.

As both sexes of the guppy are known to have an intrinsic mating preference for large body size (Reynolds and Gross 1992; Herdman et al. 2004), we took two different-sized stimulus individuals from one isolation tank and introduced them into one of the two lateral compartments each.

Stimulus individuals were exchanged between subsequent trials. Afterward, a focal fish of the opposite sex was introduced into a clear Plexiglas cylinder (10 cm diameter), placed centrally in the neutral zone, and was left undisturbed for 5 min. After this habituation period, we gently lifted the cylinder and measured the amount of time the focal fish spent in each of the two preference zones, i.e., near the large and small stimulus fish during a 5-min observation period. To account for potential side biases, the focal fish was then gently placed back into the cylinder and stimulus fish were interchanged. After 5 min of habituation, measurement was repeated for another 5 min. Like in previous studies, we decided *a priori* to discard trials in which focal fish spent more than 80% of their choice time during both parts of the experiment in the same compartment as side biases (e.g., Plath et al. 2004), but no cases of side-bias were observed. This initial preference test consisting of two measurements is henceforth called the first part of a trial (Fig. 1a).

Times spent near each stimulus fish during the two test units were summed, and if the focal fish spent more than 66% of the time with one of the stimuli they were considered to have shown a preference and were used for the subsequent test phases. Trials in which focal fish did not show a strong preference were terminated at this point (Schlupp and Ryan 1997). Out of 81 initially tested wild-type males, 15 did not show a preference for one of the two stimulus females; similarly, 17 out of 85 females showed no preference. In the domestic stock, 17 out of 61 initially tested males did not show a preference and out of 58 females, 16 tests were terminated after the first part of a trial.

Directly after the first part, the focal fish was retransferred into the Plexiglas cylinder in the center of the neutral zone ("observation phase"; Fig. 1b). We created situations during which the focal fish either had no opportunity to copy (control), or it could observe visual interactions and associations between the model individual and the formerly non-preferred mate, or physical (i.e., sexual) interactions of the latter two. To do so, focal fish were randomly assigned to one of the following three treatments: In the control treatment (1), an empty transparent Plexiglas cylinder was placed in the front portion of the non-preferred individual's compartment. To control for any changes in the behavior of the stimulus fish due to this procedure, we introduced another Plexiglas cylinder also into the compartment of the formerly preferred individual but covered this portion of the stimulus compartment by an opaque Plexiglas sheet and thus prevented the focal fish from seeing it (Fig. 1b).

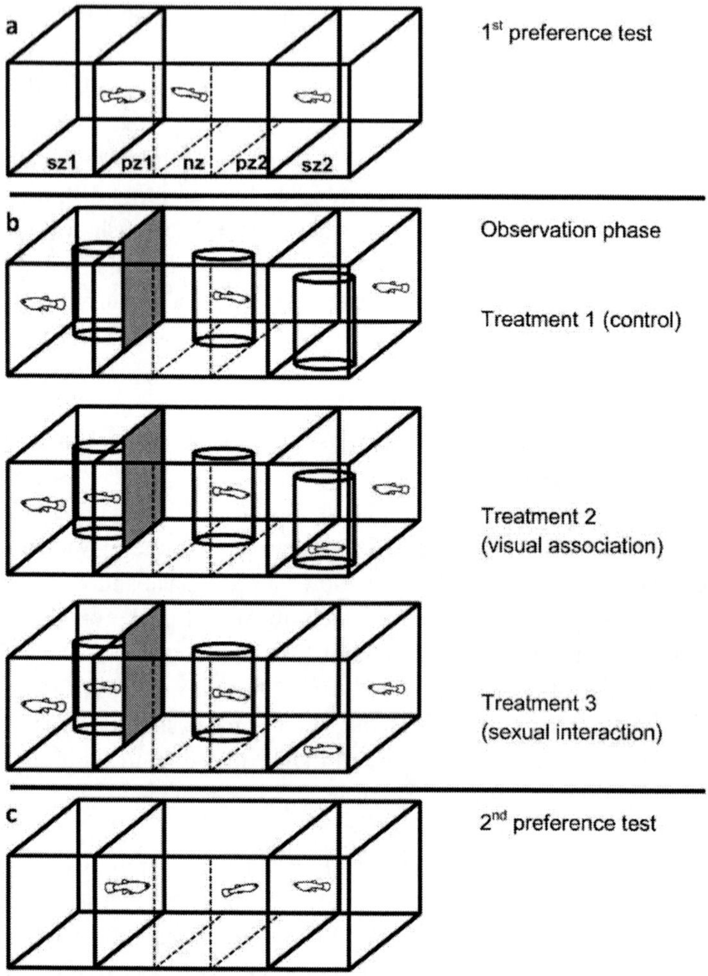

Figure 1. Experimental setup used to examine mate choice copying in *P. reticulata*. In the first preference test (a), we measured the time the focal individual spent in the preference zones (pz1 or pz2) near either stimulus fish (sz1 or sz2). During the observation phase (b), focal fish were randomly assigned to one of three treatments: treatment 1, control without a model individual; treatment 2: visual association, where the focal fish could observe a model individual visually interacting with the formerly non-preferred stimulus, and treatment 3: sexual interactions, during which they could observe physical interactions between the model individual and the formerly non-preferred stimulus. The second preference test (c) was analogous to the first and enabled us to compare preferences before and after the focal fish had the possibility to copy the model's mate choice. Depicted is the test scheme for male mate choice copying; tests for female mate choice copying were conducted analogously.

Treatment (2) was similar to previous mate copying tests in *P. latipinna* (Schlupp and Ryan 1997). Again, two Plexiglas cylinders were placed in the front portion of both compartments holding the stimulus fish (see above), but we now introduced one model individual per cylinder, so the focal fish could observe a model of the same sex visually interact with the formerly non-preferred mate, but was prevented from seeing the other model interact with the formerly preferred mate.

To allow physical interactions between the model individual and the formerly non-preferred stimulus, we introduced the model individual without a Plexiglas cylinder into the partition of the formerly non-preferred mate [treatment (3)]. Hence, the model individual could interact sexually with the stimulus fish. The previously preferred mate was able to interact visually with a fish of the opposite sex in a Plexiglas cylinder just like in treatment (2) (the cylinder with this fish was again hidden behind an opaque plastic sheet as described before). We quantified pre-copulatory behavior (nipping, mean ± SE, wild-type 10 ± 8; domestic 8 ± 10) and copulations (gonopodial thrusting, wild-type 3 ± 5; domestic 1 ± 3) in the stimulus-model pair during the 20 min of observation. After the 20-min observation phase, the model individuals were removed and measurement of association preferences (including switching of side assignments of the stimulus fish between the two test units) was carried out as described for the first part of the experiment (Fig. 1c). After the tests, the standard lengths of all fish were measured to the nearest millimeter (Table 1a).

Table 1. Body size of the test fish. Given are mean (± SE) standard lengths (SL in mm) of the focal, stimulus (large and small) as well as model fish ('preferred' = model interacted with preferred stimulus; 'non-preferred' = model interacted with non-preferred stimulus)

Population/Sex	focal	large	small	model preferred	model non-preferred
a) Experiment 1					
Domestic, females (*N* = 42)	33.0 ± 0.7	27.5 ± 0.3	21.5 ± 0.2	33.6 ± 1.0	34.4 ± 0.7
Domestic, males (*N* = 44)	22.7 ± 0.4	35.5 ± 0.5	23.8 ± 0.4	23.3 ± 0.4	22.9 ± 0.4
Wild-type, females (*N* = 68)	28.9 ± 1.1	24.8 ± 0.6	18.2 ± 0.4	27.3 ± 1.3	28.3 ± 1.3
Wild-type, males (*N* = 66)	23.7 ± 0.3	36.4 ± 0.4	29.4 ± 0.6	24.3 ± 0.5	24.6 ± 0.6
b) Experiment 2					
Wild-type, females (*N* = 21)	28.0 ± 2.8				
Wild-type, males (*N* = 22)	21.0 ± 1.5				

Experiment 2

To test if males and females have a predilection for either of the two copying situations, we used dichotomous association preference tests in conjunction with video playback techniques (Makowicz et al. 2010; Bierbach et al. 2013b). We created video sequences showing (1) a male and a female separated by an (invisible) PVC partition or (2) a pair of fish allowed to interact freely. We used only wild-type fish in this experiment. Males for the production of video recordings were isolated for one week in 10-L tanks to ensure that they were motivated to mate. The recording tank was divided into two same-sized compartments by an opaque PVC partition and a camcorder (Panasonic HDC-TM60EG-K) was set up 9 cm from the front wall to capture both compartments simultaneously. For the videos showing 'visual association', one female (SL ± SD, 29.2 ± 0.9 mm) and one male (22.6 ± 0.7 mm) were placed in either compartment, and after a 5-min acclimatization phase were filmed for at least 6 minutes. For videos showing direct interactions, we removed the PVC-partition after the acclimatization phase so males could mate with the female. We recorded numbers of nipping (8 ± 8) and mating attempts (8 ± 5). For each treatment and prepared 10 videos (resolution 800 x 600 px, saved as avi-files) that were later displayed via MS Powerpoint (Microsoft Office 2010).

For the binary association preference tests we followed the protocol of Bierbach et al. (2011a). The test tank (60 x 30 x 30 cm) was visually divided into three sections: two lateral preference zones of 10 cm width and a central, neutral zone of 40 cm width. The tank was filled with aged tap water to a height of 25 cm. Water was heated and aerated between trials, but heater and air stone were removed prior to testing. The bottom of the test tank was covered with fine white quartz sand. Illumination was provided by two 60 W fluorescent lamps on the ceiling of the test room. We placed two identical video screens (Samsung SyncMaster P2470LHD, 24 in.; resolution 1,280 × 768 pixels) at both short sides of the test tank and covered all remaining parts of the test tank's walls (including the front wall) with grey cardboard to reduce disturbances from the outside. The test fish was observed from above by use of a mirror.

To initiate a trial, we introduced a focus fish into a transparent Plexiglas cylinder placed centrally in the neutral zone and gave the test subject 5 min for acclimatization, during which the monitors showed a uniformly gray display. Afterwards, we released the focal fish and started the video playback, showing the two types of video sequences (see above) on the left and right side (side-assignment was altered between trials). We measured the time the focal individual spent in each preference zone for 5 min. Focal fish were then reintroduced into the Plexiglas cylinder and given another 5 min of habituation while both screens displayed gray background only. Subsequently, measurement of association times was repeated for another 5 min with sides of the video presentation interchanged. We summed up association times from both parts and calculated the strength of preference (SOP) as follows:

$$\mathrm{SOP} = \frac{t_{sex\ int} - t_{vis\ assoc}}{t_{sex\ int} + t_{vis\ assoc}}$$

where $t_{sex\ int}$ is the time spent in the compartment next to the sexual interaction-sequence and $t_{vis\ assoc}$ the time spent next to the visual association-sequence. Upon completion of a trial, we measured SL of the focal fish to the nearest millimeter (Table 1b).

Statistical Analysis

All data were tested for normal distribution using Kolmogorov-Smirnov tests and were presented as mean ± standard error (SE) throughout. Statistical analyses were conducted using IBM SPSS 18.

Experiment 1

In a first step, we evaluated the focal individuals' preference for large body size. We compared the amount of time focal fish spent near large and small stimulus individuals during the initial preference test (first part) and during the second part of the tests for all three treatments separately, using paired sample *t*-tests.

Our main question was whether focal fish would alter their individual mate choice decisions after they had observed a model individual visually [treatment (2)] or sexually [treatment (3)] interact with the formerly non-preferred stimulus. We, therefore, calculated a score expressing the change of mating preference as the difference between individuals' relative association time with the initially non-preferred female during the second part and relative association time near the same female during the first part ('copying score'; see Heubel et al. 2008; Bierbach et al. 2011a, 2013b), such that no change would lead to a score of zero, and positive values would indicate a relative increase in time spent near the initially non-preferred stimulus due to copying.

Scores were compared using a full-factorial univariate General Linear Model (GLM) with 'treatment', 'sex' and 'population' as fixed factors. As mate choice copying in other poeciliids depends on the size difference between the two stimulus individuals (Witte and Ryan 1998), we initially included stimulus body size difference (SL large – SL small) and focal individuals' body size as covariates but removed them as neither the covariates themselves nor any of their interactions had a significant effect (body size difference: $F_{1, 198} = 2.11$, $P = 0.15$; focal body size: $F_{1, 198} = 0.22$, $P = 0.64$). We also removed the non-significant 3-way interaction of all main factors from the final model ($F_{2, 208} = 2.55$, $P = 0.081$) as well as non-significant 2-way interactions (treatment × sex: $F_{2,209} = 2.1$, $P = 0.16$; sex × population: $F_{1,210} = 0.59$, $P = 0.44$). We used Fisher's LSD tests to *post hoc* analyze differences between treatments.

To test for a potential effect of body size differences between model and focal individuals on the strength of mate choice copying (copying score), we included body size difference (SL focal – SL model) as a covariate in another GLM while analyzing only the subset of data from treatments (2) and (3).

To evaluate whether the number of sexual interactions that a focal fish observed during the observation phase affected the strength of copying in treatment 3, we included numbers of nippings and mating attempts as covariates in another GLM analyzing the subset of data from treatment (3).

Experiment 2

We compared times males and females spent near the video sequences showing sexual interactions and visual associations for each sex separately using paired *t*-tests. Furthermore, we compared the strength of preference (SOP) between males and females in a GLM including 'video ID' as random factor and focal individuals' body size (SL) as a covariate.

RESULTS

Experiment 1: Mate Choice Copying in Males and Females

Preference for Large Mating Partners

Males and females of both populations spent significantly more time in association with larger stimulus fish during the first part of the preference tests (Fig. 2). This preference was largely maintained during the second part of the tests (i.e. after the observation phase) with two notable exceptions: females of both populations ceased expressing a preference for the larger stimulus male in treatment (2) (visual association) and spent almost equal amounts of time near the two stimuli. Also, no significant preference was seen in males from the domestic stock in treatment (3) (sexual interaction), but qualitatively males still tended to spent more time associating with the larger female ($P = 0.11$; Fig. 2).

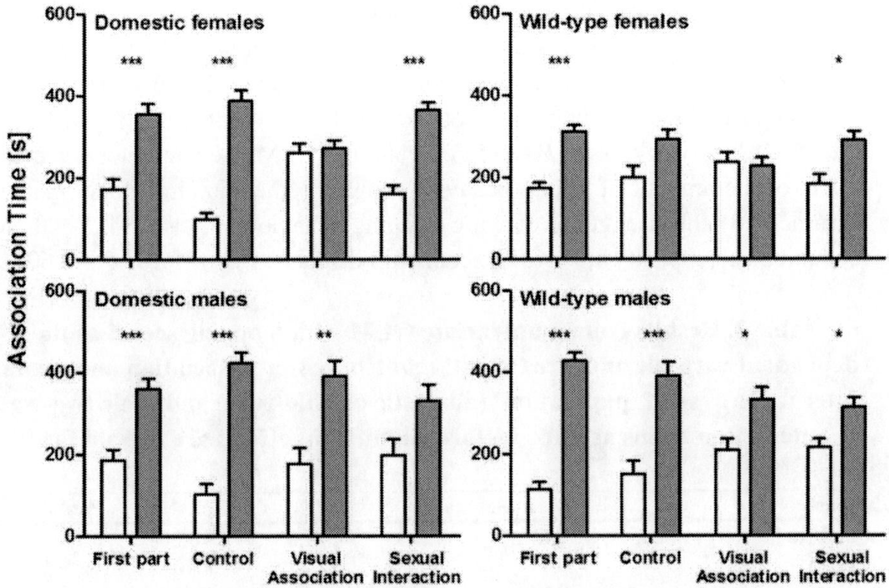

Figure 2. Time focal fish spent in association with the large and the small stimulus individuals during the first part of the choice tests (left) and during the second part in the three copying treatments. During the observation phase preceding the second part of the choice tests either no copying was possible (control), or a model fish could interact visually or physically with the previously non-preferred stimulus. Depicted are association times (± SE); significant *P*-values are from paired *t*-tests (*$P < 0.05$; **$P < 0.01$; ***$P < 0.001$).

Changes in Individual Preferences

The final GLM detected a significant effect of the main factor 'treatment' and *post hoc* LSD tests indicate that, overall, test subjects showed consistent mate choice in the control treatment (copying scores close to zero in Fig. 3; LSD: $P < 0.001$ compared to both other treatments), but copied to a similar extent in treatment (2) (positive copying score for 'visual association') and treatment (3) [positive copying score for 'sexual interaction'; LSD test, treatment (2) vs. treatment (3): $P = 0.52$]. Moreover, a significant interaction effect of 'treatment by population' was uncovered, suggesting that the two populations differed in their responses to the different copying treatments, even though effect size (estimated as partial eta²) was lower for the interaction term compared to the main effect 'treatment' (Table 2). Nevertheless, estimated marginal means derived from GLM revealed that domesticated guppies, overall, copied more than wild-type guppies (Fig. 3). Sexes did not differ statistically in their copying behavior (Table 2).

We also tested for an effect of the focal and model individuals' body size difference on the strength of copying within the subset of data from treatments (2) and (3). However, body size difference had no statistically significant effect (GLM; $F_{1,142} = 0.40$, $P = 0.53$). In another GLM, we could not detect an effect of the amount of sexual behavior shown by the stimulus-model pair in treatment (3) and strength of copying (nipping behavior: $F_{1,64} = 0.85$, $P = 0.36$; copulations: $F_{1,64} = 0.18$, $P = 0.68$; courtship displays: $F_{1,64} = 0.27$, $P = 0.60$).

Table 2. Results from a univariate GLM with 'copying score' as the dependent variable and 'treatment' (control, visual association and sexual interaction), 'sex', 'population' (domestic or wild-type) and their two-way interaction terms as fixed factors. Significant effects are in bold face

Source	Mean Square	*df*	*F*	*P*	Partial eta²
Treatment	**1.67**	**2**	**19.63**	**<0.001**	**0.17**
Sex	0.17	1	2.17	0.16	<0.01
Population	0.04	1	0.29	0.64	<0.01
Treatment * Population	**0.41**	**2**	**4.56**	**0.01**	**0.05**
Error	0.09	213			

Experiment 2: Preference for Different Copying Situations

Neither females nor males showed a preference for either type of video (males, visual association: 194.7 ± 17.3 s; sexual interactions: 172.6 ± 16.6 s; paired *t*-test: $t_{21} = -0.91$, $P = 0.37$; females, visual association: 205.5 ± 20.6 s; sexual interactions 161.6 ± 22.6 s; $t_{20} = -1.16$, $P = 0.26$). Also, there was no significant effect of the factor 'sex' ($F_{1,30} = 0.14$, $P = 0.71$) when comparing SOP-values using GLM. Likewise, the covariate (focal individuals' body size) had no significant effect ($F_{1,30} = 0.20$, $P = 0.66$).

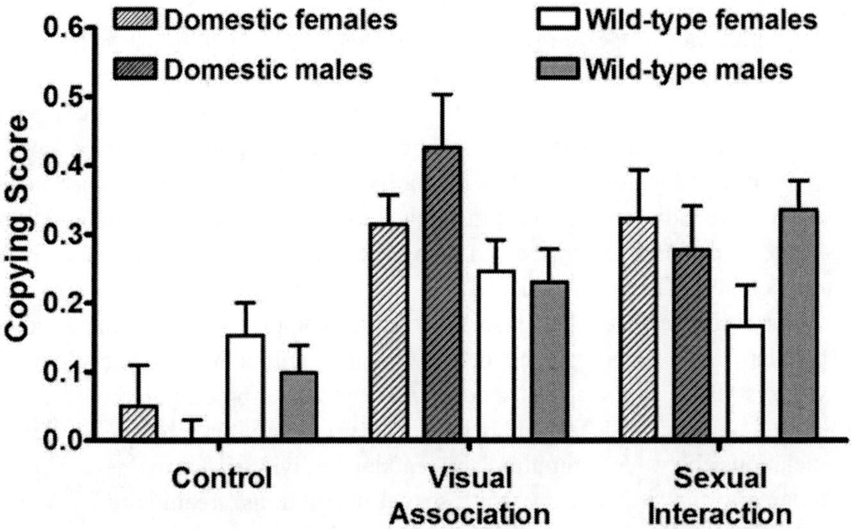

Figure 3. Changes in individual mating preferences when focal fish were given an opportunity to copy another individual's mate choice. During the observation phase preceding the second part of the choice tests either no copying was possible (control, left) or a model fish could interact visually (middle) or physically (right) with the previously non-preferred stimulus fish. Depicted are copying scores (see main text), whereby positive values indicate that preferences for the initially non-preferred stimulus individuals increased in strength.

DISCUSSION

Copying the mate choice decisions of other individuals provides benefits to both males and females, as copying helps reduce costs of mate searching (Gibson and Höglund 1992; Schlupp and Ryan 1997; Witte and Noltemeier 2002). At the same time, mate choice copying is associated with costs, namely increased sperm competition risk for males that copy other males' mate choice (see Evans and Magurran 1999) and increased sexual harassment for copying females (see Plath et al. 2007). Cost-benefit trade-offs are predicted to shape mate choice decisions in general (Kokko et al. 2003), and similar trade-offs ought to affect copying behavior as well.

Guppies have a promiscuous mating system characterized by strong mating competition among males, leading to high levels of sperm competition and sexual harassment (Magurran and Seghers 1994; Evans and Magurran 1999; Plath et al. 2007), and so the costs associated with mate choice copying should be high. Our current study was motivated by the hypothesis that individuals copy opportunistically when costs associated with mate choice copying are moderate or low (i.e., in the visual association treatment), but cease copying other individuals' mate choice decisions when costs are high (in the direct interaction treatment). Support for this hypothesis comes from a study that found males of the Atlantic molly (*P. mexicana*), another member of the family Poeciliidae, to reduce mate choice copying when sperm competition risk was experimentally increased using the same experimental approach we emplyoed here (Bierbach et al. 2011a).

Our study revealed that male and female guppies copy the mate choice decisions of conspecifics, and there was no significant difference between sexes in strength of mate choice copying. Contrary to prediction, however, we found no significant difference between the copying situation involving mere associations between stimulus and model individuals [treatment (2)] as compared to the situation involving sexual interactions [treatment (3)]. We do not have a compelling explanation at hand for the apparent differences between Atlantic mollies (Bierbach et al. 2011a) and guppies (this study). It seems tempting to argue though that copying provides greater benefits to the copier in guppies than Atlantic mollies, such that male and female guppies are willing to accept costs arising from sperm competition and harassment and thus, copy in a broader range of social contexts. Future studies should make an attempt to shed light on this question. For example, video animation techniques could be employed to create virtual copying situations that gradually increase the amount of sexual interactions between stimulus and model individuals. Comparing Atlantic mollies and guppies in such an experimental design would be an elegant test to our new hypothesis, as Atlantic mollies are predicted to cease copying at a lower threshold value of sexual behavior. This experiment, of course, still leaves open the important question of *why* species of poeciliids might differ in their propensity to copy the mate choice of others. Comparative approaches using different species and genera of poeciliid fishes might shed light on the question of how differences in social organization, mate encounter rate, male aggression, and other factors, affect the evolution of copying behavior.

Beside the strong treatment effect, we also found a weak but statistically significant interaction effect of 'population by treatment', indicating differences in the magnitude of mate copying between populations, and domestic guppies copied more than the descendants of wild-caught animals. Noteworthy, several studies reported on a lack of female mate choice copying behavior in feral and wild-type pet shop guppies (Brooks 1996, 1999; Lafleur et al. 1997), resulting in some controversy about the importance of mate choice copying for sexual selection (Dugatkin 1998). The majority of studies suggest that female mate choice copying is a universal and widespread behavior—at least in guppies (Dugatkin 1992, 1996; Dugatkin and Godin 1992, 1993; Godin & Hair 2009)—while Brooks (1998) argued that population differences may exist. Our results indeed suggest some variation in mate choice copying among populations. We have no compelling explanation at hand as to the question of why domestic guppies copied more than wild type guppies. The exact founder population(s) from which the artificial strain was derived is unknown, so it remains possible that the ancestors of our domestic strain already exhibited stronger copying; likewise, we cannot rule out the possibility that the domestication process has selected for increased copying, even though the latter scenario seems unlikely.

ACKNOWLEDGMENTS

We like to thank Holger Geupel for help with animal care. Financial support came from the German Research Foundation (Deutsche Forschungsgemeinschaft, DFG) to MP (PL 470/3-1).

REFERENCES

Andersson MB (1994) *Sexual selection*. Princeton University Press, Princeton NJ

Applebaum SL, Cruz A (2000) The Role of Mate-Choice Copying and Disruption Effects in Mate Preference Determination of *Limia perugiae* (Cyprinodontiformes, Poeciliidae). *Ethology* 106 (10):933–944

Aquiloni L, Buřič M, Gherardi F (2008) Crayfish females eavesdrop on fighting males before choosing the dominant mate. *Curr Biol* 18 (11):R462

Auld HL, Punzalan D, Godin JJ, Rundle HD (2009) Do female fruit flies (*Drosophila serrata*) copy the mate choice of others? *Behavioural Processes* 82 (1):78–80

Becher SA, Magurran AE (2004) Multiple mating and reproductive skew in Trinidadian guppies. *P Roy Soc B-Biol Sci* 271 (1543):1009–1014

Bierbach D, Girndt A, Hamfler S, Klein M, Mücksch F, Penshorn M, Schwinn M, Zimmer C, Schlupp I, Streit B, Plath M (2011b) Male fish use prior knowledge about rivals to adjust their mate choice. *Biology Letters* 7 (3):349–351

Bierbach D, Jung CT, Hornung S, Streit B, Plath M (2013b) Homosexual behaviour increases male attractiveness to females. *Biology Letters* 9 (1):20121038

Bierbach D, Kronmarck C, Hennige-Schulz C, Stadler S, Plath M (2011a) Sperm competition risk affects male mate choice copying. *Behav Ecol Sociobiol* 65 (9):1699–1707

Bierbach D, Sassmannshausen V, Streit B, Arias-Rodriguez L, Plath M (2013a) Females prefer males with superior fighting abilities but avoid sexually harassing winners when eavesdropping on male fights. *Behav Ecol Sociobiol* 67 (4):675–683

Bisazza A, Marin G (1995) Sexual selection and sexual size dimorphism in the eastern mosquitofish *Gambusia holbrooki* (Pisces Poeciliidae). *Ethol Ecol Evol* 7 (2):169–183

Bonnie KE, Earley RL (2007) Expanding the scope for social information use. *Anim Behav* 74 (2):171–181

Briggs SE, Godin JJ, Dugatkin LA (1996) Mate-choice copying under predation risk in the Trinigadian guppy (*Poecilia reticulata*). *Behavioral Ecology* 7 (2):151–157

Brooks R (1996) Copying and the repeatability of mate choice. *Behav Ecol Sociobiol* 39 (5):323–329

Brooks R (1998) The importance of mate copying and cultural inheritance of mating preferences. *Trends in Ecology & Evolution* 13 (2):45–46

Brooks R (1999) Mate choice copying in guppies: females avoid the place where they saw courtship. *Behaviour* 136 (4):411–421

Candolin U (2003) The use of multiple cues in mate choice. *Biol Rev* 78 (4):575–595

Clutton-Brock T, McComb K (1993) Experimental tests of copying and mate choice in fallow deer (*Dama dama*). *Behav Ecol* 4 (3):191–193

Constantz GD (1984) Sperm competition in Poeciliid fishes. In: Smith RL (ed) *Sperm Competition and the Evolution of Animal Mating systems*. Elsevier Science, Oxford, pp 465–485

Constantz GD (1989) Reproductive biology in poeciliid fishes. In: Meffe GK (ed) *Ecology and evolution of livebearing fishes* (Poeciliidae). Prentice Hall, Englewood Cliffs, NJ, pp 33–50

Dabelsteen T (2005) Public, private or anonymous? Facilitating and countering eavesdropping. In: McGregor PK (ed) *Animal communication networks*. Cambridge University Press, Cambridge, pp 38–62

Danchin É (2004) *Public Information: From Nosy Neighbors to Cultural Evolution. Science* 305 (5683):487–491

Darrow KO, Harris WA (2004) Characterization and Development of Courtship in Zebrafish, *Danio rerio*. Zebrafish 1 (1):40–45

Darwin C (1871) *The descent of man, and Selection in relation to sex*. J. Murray, London

Dosen LD, Montgomerie R (2004) Mate preferences by male guppies (*Poecilia reticulata*) in relation to the risk of sperm competition. *Behav Ecol Sociobiol* 55 (3):266–271

Druen M, Dugatkin LA (2011) Communication networks. In: Evans JP, Pilastro A, Schlupp I (eds) *Ecology and evolution of poeciliid fishes*. University of Chicago Press, Chicago, IL, pp 84–113

Dugatkin LA (1992) Sexual Selection and Imitation: Females Copy the Mate Choice of Others. *Am Nat* 139 (6):1384–1389

Dugatkin LA (1996) Copying and Mate Choice. In: Heyes CM, Galef BG, jr (eds) *Social learning in animals: the roots of culture*. Academic Press, New York, pp 85–105

Dugatkin LA (1998) A comment on Lafleur et al.'s re-evaluation of mate-choice copying in guppies. *Anim Behav* 56:513–514

Dugatkin LA (1998) Genes, copying, and female mate choice: shifting thresholds. *Behav Ecol* 9 (4):323–327

Dugatkin LA, Godin JJ (1992) Reversal of Female Mate Choice by Copying in the Guppy (*Poecilia reticulata*). *P Roy Soc B-Biol Sci* 249 (1325):179–184

Dugatkin LA, Godin JJ (1993) Female mate copying in the guppy (*Poecilia reticulata*): age-dependent effects. *Behav Ecol* 4 (4):289–292

Evans JP, Magurran AE (1999) Male mating behaviour and sperm production characteristics under varying sperm competition risk in guppies. *Anim Behav* 58 (5):1001–1006

Evans JP, Magurran AE (2001) Patterns of sperm precedence and predictors of paternity in the Trinidadian guppy. *P Roy Soc B-Biol Sci* 268 (1468):719–724

Evans JP, Pilastro A (2011) Postcopulatory sexual selection. In: Evans JP, Pilastro A, Schlupp I (eds) *Ecology and evolution of poeciliid fishes*. University of Chicago Press, Chicago, IL, pp 197–208

Farr JA (1989) Sexual selection and secondary sexual differentiation in poeciliids: determinants of male mating success and the evolution of female choice. In: Meffe GK (ed) *Ecology and evolution of livebearing fishes* (Poeciliidae). Prentice Hall, Englewood Cliffs, NJ, pp 91–123

Fitzsimmons LP, Foote JR, Ratcliffe LM, Mennill DJ (2008) Eavesdropping and communication networks revealed through playback and an acoustic location system. *Behav Ecol* 19 (4):824–829

Freed-Brown G, White DJ (2009) Acoustic mate copying: female cowbirds attend to other females' vocalizations to modify their song preferences. *P Roy Soc B-Biol Sci* 276 (1671):3319–3325

Frommen JG, Rahn AK, Schroth SH, Waltschyk N, Bakker TCM (2009) Mate-choice copying when both sexes face high costs of reproduction. *Evol Ecol* 23 (3):435–446

Galef BG, jr, Lim TC, Gilbert GS (2008) Evidence of mate choice copying in Norway rats, *Rattus norvegicus. Anim Behav* 75 (3):1117–1123

Galef BG, jr, White DJ (1998) Mate-choice copying in Japanese quail, *Coturnix coturnix japonica. Anim Behav* 55 (3):545–552

Gibson RM, Höglund J (1992) Copying and sexual selection. *Trends Ecol Evol* 7 (7):229–232

Godin JJ, Hair K (2009) Mate-choice copying in free-ranging Trinidadian guppies (*Poecilia reticulata*). *Behaviour* 146 (10):1443–1461

Grant JWA, Green LD (1996) Mate copying versus preference for actively courting males by female Japanese medaka (*Oryzias latipes*). *Behav Ecol* 7 (2):165–167

Heubel KU, Hornhardt K, Ollmann T, Parzefall J, Ryan MJ, Schlupp I (2008) Geographic variation in female mate-copying in the species complex of a unisexual fish, *Poecilia formosa. Behaviour* 145 (8):1041–1064

Höglund J, Lundberg A, Alatalo RV (1990) Copying the Mate Choice of Others? Observations On Female Black Grouse. *Behaviour* 114 (1):221–231

Houde AE (1997) *Sex, color, and mate choise in guppies*. Princeton University Press, Princeton NJ

Jeswiet SB, Lee-Jenkins SSY, Godin JJ (2011) Concurrent effects of sperm competition and female quality on male mate choice in the Trinidadian guppy (*Poecilia reticulata*). *Behav Ecol* 23 (1):195–200

Köhler A, Hildenbrand P, Schleucher E, Riesch R, Arias-Rodriguez L, Streit B, Plath M (2011) Effects of male sexual harassment on female time budgets, feeding behavior, and metabolic rates in a tropical livebearing fish (*Poecilia mexicana). Behav Ecol Sociobiol* 65 (8):1513–1523

Lafleur DL, Lozano GA, Sclafani M (1997) Female mate-choice copying in guppies, *Poecilia reticulata*: a re-evaluation. *Anim Behav* 54 (3):579–586

Liley NR (1996) Ethological isolating mechanisms in four sympartic species of Poeciliid fishes. *Behaviour* 13 (suppl.) (1):197

Magurran AE (2001) Sexual conflict and evolution in Trinidadian guppies. *Genetica* 112:463–474

Magurran AE (2005) *Evolutionary ecology: the Trinidadian guppy*. Oxford University Press, Oxford, UK

Magurran AE (2011) Sexual coercion. In: Evans JP, Pilastro A, Schlupp I (eds) *Ecology and evolution of poeciliid fishes.* University of Chicago Press, Chicago, IL, pp 209–217

Magurran AE, Seghers BH (1994) A Cost of Sexual Harassment in the Guppy, *Poecilia reticulata. P Roy Soc B-Biol Sci* 258 (1351):89–92

Makowicz A, Plath M, Schlupp I (2010) Using video playback to study the effect of an audience on male mating behavior in the Sailfin molly (*Poecilia latipinna*). *Behavioural Processes* 85 (1):36–41

McGregor PK, Peake TM (2000) Communication networks: social environments for receiving and signalling behaviour. *Acta Ethol* 2 (2):71–81

Mery F, Varela SA, Danchin É, Blanchet S, Parejo D, Coolen I, Wagner RH (2009) Public Versus Personal Information for Mate Copying in an Invertebrate. *Curr Biol* 19 (9):730–734

Naguib M, Amrhein V, Kunc HP (2004) Effects of territorial intrusions on eavesdropping neighbors: communication networks in nightingales. *Behav Ecol* 15:1011–1015

Nordell SE, Valone TJ (1998) Mate choice copying as public information. *Ecol Lett* 1 (2):74–76

Ophir AG, Galef BG, jr (2003) Female Japanese quail that 'eavesdrop' on fighting males prefer losers to winners. *Anim Behav* 66 (2):399–407

Ophir AG, Galef BG, jr (2004) Sexual experience can affect use of public information in mate choice. *Anim Behav* 68:1221–1227

Parker GA (1979) Sexual selection and sexual conflict. In: Blum MS, Blum NA (eds) *Sexual selection and reproductive competition in insects*. Academic Press, New York, San Francisco, London, pp 123–166

Parzefall J (1969) Zur vergleichenden Ethologie verschiedener *Mollienesia*-Arten einschliesslich einer Höhlenform vom *M. sphenops. Behaviour* 33 (1/2):1–37

Peake TM (2005) Eavesdropping in communication networks. In: McGregor PK (ed) *Animal communication networks*. Cambridge University Press, Cambridge, pp 13–37

Pilastro A, Gasparini C, Boschetto C, Evans JP (2007) Colorful male guppies do not provide females with fecundity benefits. *Behav Ecol* 19 (2):374–381

Plath M, Makowicz AM, Schlupp I, Tobler M (2007) Sexual harassment in live-bearing fishes (Poeciliidae): comparing courting and noncourting species. *Behav Ecol* 18 (4):680–688

Plath M, Parzefall J, Körner KE, Schlupp I (2004) Sexual selection in darkness? Female mating preferences in surface- and cave-dwelling Atlantic mollies, *Poecilia mexicana* (Poeciliidae, Teleostei*). Behav Ecol Sociobiol* 55 (6):596–601

Plath M, Parzefall J, Schlupp I (2003) The role of sexual harassment in cave and surface dwelling populations of the Atlantic molly, *Poecilia mexicana* (Poeciliidae, Teleostei). *Behav Ecol Sociobiol* 54 (3):303–309

Price AC, Weadick CJ, Shim J, Rodd FH (2008) Pigments, Patterns, and Fish Behavior. *Zebrafish* 5 (4):297–307

Pruett-Jones S (1992) Independent versus nonindependent mate choice: do females copy each other? *Am Nat* 140 (6):1000–1009

Reynolds JD, Gross MR (1992) Female Mate Preference Enhances Offspring Growth and Reproduction in a Fish, *Poecilia reticulata. P Roy Soc B-Biol Sci* 250 (1327):57–62

Reznick D, Endler JA (1982) The Impact of Predation on Life History Evolution in Trinidadian Guppies (*Poecilia reticulata*). *Evolution* 36:160–177

Rosenthal GG (2007) Spatiotemporal Dimensions of Visual Signals in Animal Communication. *Annu. Rev. Ecol. Evol. Syst.* 38 (1):155–178

Rosenthal HL (1952) Observations on Reproduction of the Poeciliid *Lebistes reticulatus* (Peters). *Biological Bulletin* 102 (1):30

Schlupp I, Marler C, Ryan MJ (1994) Benefit to male sailfin mollies of mating with heterospecific females. *Science* 263 (5145):373–374

Schlupp I, Ryan MJ (1997) Male Sailfin mollies (*Poecilia latipinna*) copy the mate choice of other males. *Behav Ecol* 8 (1):104–107

Swaddle JP, Cathey MG, Correll M, Hodkinson BP (2005) Socially transmitted mate preferences in a monogamous bird: a non-genetic mechanism of sexual selection. *P Roy Soc B-Biol Sci* 272 (1567):1053–1058

Tobler M, Schlupp I, Plath M (2011) Costly interactions between the sexes: combined effects of male sexual harassment and female choice? *Behav Ecol* 22 (4):723–729

Uller T, Johansson LC (2003) Human mate choice and the wedding ring effect. *Hum Nat* 14 (3):267–276

Valone TJ (2007) From eavesdropping on performance to copying the behavior of others: a review of public information use. *Behav Ecol Sociobiol* 62 (1):1–14

Valone TJ, Templeton JJ (2002) Public information for the assessment of quality: a widespread social phenomenon. *Philos T Roy Soc B* 357 (1427):1549–1557

Warner RR, Wernerus F, Lejeune P, van den Berghe E (1995) Dynamics of female choice for parental care in a fish species where care is facultative. *Behav Ecol* 6 (1):73–81

Waynforth D (2007) Mate Choice Copying in Humans. Hum Nat 18 (3):264–271

Westneat DF, Walters A, McCarthy TM, Hatch MI, Hein WK (2000) Alternative mechanisms of nonindependent mate choice. *Anim Behav* 59 (3):467–476

White DJ, Galef BG, jr (1999) Mate choice copying and conspecific cueing in Japanese quail, *Coturnix coturnix japonica. Anim Behav* 57 (2):465–473

White DJ, Galef BG, jr (2000) Social effects on mate choices of male Japanese quail, *Coturnix japonica*. Anim Behav 59:975–979

Widemo MS (2005) Male but not female pipefish copy mate choice. *Behavioral Ecology* 17 (2):255–259

Witte K, Nöbel S (2011) Learning and Mate Choice. In: Brown C, Laland K, Krause J (eds) *Fish cognition and behavior*, 2nd edn. Wiley-Blackwell, Ames, IA, pp 81–107

Witte K, Noltemeier B (2002) The role of information in mate-choice copying in female sailfin mollies (*Poecilia latipinna). Behav Ecol Sociobiol* 52 (3):194–202

Witte K, Ryan MJ (1998) Male body length influences mate-choice copying in the sailfin molly *Poecilia latipinna*. Behav Ecol 9 (5):534–539

Witte K, Ryan MJ (2002) Mate choice copying in the sailfin molly, *Poecilia latipinna*, in the wild. *Anim Behav* 63 (5):943–949

Witte K, Ueding K (2003) Sailfin molly females (*Poecilia latipinna*) copy the rejection of a male. *Behav Ecol* 14 (3):389–395

Wong BBM, McCarthy M (2009) Prudent male mate choice under perceived sperm competition risk in the eastern mosquito fish. *Behav Ecol* 20 (2):278–282

Yorzinski JL, Platt ML, Reby D (2010) Same-Sex Gaze Attraction Influences Mate-Choice Copying in Humans. *PLoS ONE* 5 (2):e9115

Ziege M, Hennige-Schulz C, Mücksch F, Bierbach D, Tiedemann R, Streit B, Plath M (2012) A comparison of two methods to assess audience-induced changes in male mate choice. *Current Zoology* 58 (1):84–94

In: Sexual Selection
Editors: R. Geldani and M. Davin
ISBN: 978-1-62808-805-2

Chapter 4

SEXUAL SELECTION UNDER PARENTAL CHOICE: MATING STRATEGIES AND REPRODUCTIVE SUCCESS

Menelaos Apostolou and Marianna Zacharia
University of Nicosia, Nicosia, Cyprus

ABSTRACT

Across pre-industrial human societies mating is regulated, with arranged marriage, in which male parents select spouses for their female relatives, being the primary mode of long-term mating. This chapter reviews the model of sexual selection under parental choice that offers a good account for these patterns of human mating. This model postulates that parent-offspring conflict over mating induces parents to control the mate choices of their children, and the spouses they select for them are individuals who conform best to their preferences. By doing so, parents become a significant evolutionary force affecting the course of sexual selection. The model is applied in order to achieve a comprehension of the evolution of specific mating strategies. In particular, it is argued that in a context where mate choice is regulated, at least three mating strategies can thrive: addressing parental choice, addressing female choice and circumventing parental and female choice by force.

INTRODUCTION

When Darwin introduced his theory of sexual selection, which postulates that certain traits have evolved to permit individuals to attract and retain mates, he identified female choice to be a primary sexual selection force (Darwin, 1871). In particular, he argued that since it is men who strive to gain access to women and not the other way round, traits which make a man to be more attractive to women are selected and increase in frequency in the population.

Darwin did not provide an explanation of why it is typically men who struggle to gain access to women and not the opposite. This theoretical gap that was successfully addressed and closed a hundred years later by Robert Trivers. Trivers (1972) emphasized that women devote, supply and provide more to their offspring than men which makes them the scarce reproductive resource over which men strive to gain access. Trivers' argument contributed considerably to the theory of sexual selection under female choice, as it offered more solid foundations. The aforementioned, combined with the observation that in Western societies men seem to sacrifice substantial resources to gain access to women, initiated and enhanced intense theorizing and empirical work centered on the model of sexual selection under female choice.

For example, Zahavi (1975) suggested that costly traits, such as the peacock's tail, evolved to reliably communicate male fitness to females. This theory was utilized by Zahavi & Zahavi (1997) and Miller (2000) who articulated that men have evolved a tendency to engage in costly behaviors like in sports in order to communicate their abilities to women in a cheat-proof fashion. Likewise, since sexual selection is directed by female choice, female preferences direct its course. Accordingly, Buss (2003) and other researchers examined mate preferences, with a focus on female mate preferences.

The model of female choice, although it bears some good theoretical basis and appears to apply to a Western context, it has a fundamental weakness namely, it is not consistent with the patterns of human mating found in modern and historical pre-industrial societies. In these societies, women are not free to exercise choice, as their mating decisions are regulated by their parents, typically their fathers, who select spouses for them (Apostolou, 2010b; Blood, 1972). Since the way of life over the previous two million years of human evolution more closely bears a resemblance to the way of life of modern pre-industrial societies compared to that of Western societies, it is reasonably possible that these patterns of mating portrayed mate choice during most of human evolution (Apostolou, 2010b). This delineates that the model of sexual selection under female choice is insufficient in explaining how sexual selection works in our species.

A different model has been recently put forward which better fits the mating patterns observed in pre-industrial human societies (Apostolou, 2007b, 2010b). In this model, conflicting interests over mating induce parents to regulate mate choice and choose individuals who comply best with their preferences, as spouses for their children. The purpose of this chapter is to review this model and apply it in understanding the evolution and the prevalence of male mating strategies. The fundamental constituent of the model is parent-offspring conflict over mating, which will be discussed next.

Parent-Offspring Conflict over Mating

All children's genes come from their parents, yet not all of parents' genes are inside their children. Consequently, as parents and children are not genetically identical, the traits of a mating candidate which are most beneficial and desirable to children are not essentially so for their parents. For this reason, parents and children have evolved asymmetrical preferences over traits, which offer them asymmetrical benefits (Apostolou, 2007a; 2008; Buunk, Park, & Dubbs, 2008; Trivers, 1974).

Perhaps the most characteristic case is genetic quality. The coefficient of relatedness of parents to children is 0.5, but the coefficient of relatedness of grandparents to grandchildren is half as much that is, 0.25. Consequently, the probability of a particular gene of an individual being passed into the next generation by a spouse or in-law would be 50% or 25%, respectively. This means that parents obtain less genetic benefits from a prospective mate of high genetic quality than their offspring (Apostolou, 2007a; Buunk et al., 2008). Beauty is a proxy of genetic quality (Thornhill & Gangestad, 1993), and therefore, is preferred more in a spouse than in an in-law (Apostolou, 2008; Buunk et al., 2008). Preference divergence is not limited solely to beauty, as research has demonstrated that parents and offspring have diverse preferences concerning family and religion background, and exciting personality (Apostolou, 2008; Buunk et al., 2008, Perilloux, Fleischman, & Buss, 2011).

Asymmetrical preferences lead to conflict among parents and children as a result of the trade-off nature of mating: constrained by their own mate value, parents and offspring have to make compromises with respect to a mating candidate's desirable qualities. However, since the two parties do not share identical preferences, they are going to make dissimilar compromises (Apostolou, 2008; Buunk et al., 2008). Therefore, if children solely exercise mate choice, they will compromise more than their parents would like over certain traits, like industriousness and social status, in order to get an attractive spouse. Conversely, genetic quality associated with beauty is less important to parents, thus the advantage from this trait is insufficient to compensate for the loss of other desirable traits. For that reason, the children's mate choice imposes a cost upon parents in the form of loss in desirable traits (Apostolou, 2008).

A study examining the tradeoffs of in-law and mate choice, asked parents to design an ideal spouse for their children, and their children to design an ideal spouse for themselves (Apostolou, 2011). Participants were given a budget of mate points and they were asked to allocate them across several desirable traits. The study demonstrated that children assigned fewer points in traits, such as similarity in religion and good family background in order to get more of exciting personality and beauty. Their parents, on the other hand, assigned fewer points on exciting personality and beauty in order to get more of good family background and similarity in religion.

These findings indicate that if children are left on their own to exercise mate choice, this choice will impose a cost on their parents with regard to loss in desirable traits. This stimulates and encourages parents to control their children's mate choices, which in turn has an impact on the course of sexual selection.

The Model of Parental Choice

Parent-offspring conflict over mating mandates that mate choice left in the hands of offspring does not maximize the fitness of parents. As a consequence, substantial evolutionary pressure is exercised on the parents to regulate the mating decisions of their children.

The prolonged period during which children rely on parental investment for survival and reproduction and the fact that parents and their kin are physically stronger than their children permit parents to control mate choice (Apostolou, 2007b). By doing so, they become effectively a sexual selection force, as traits that make an individual more likely to be chosen as an in-law are selected and increase in frequency in the population compared to less desirable ones.

Since females invest more in their offspring, they become the scarce reproductive resource to which males are seeking access (Trivers, 1972). Thus, by controlling this "resource", parents effectively control mating (Apostolou, 2007b). Furthermore, lenient parental control over female children possibly bears detrimental consequences, such as unwanted pregnancy (Perilloux, Fleischman, & Buss, 2008). Consequently, parental control is biased against female offspring, something that is also facilitated by the fact that daughters are physically weaker than sons. Parental choice is also asymmetrical in favor of male parents. By means of greater physical strength, exclusive use of weaponry and control of political institutions, male parents have more influence over their offspring's' mate choices than female parents (Apostolou, 2007b).

Nevertheless, parental control over mating is not absolute. Parents cannot always effectively "protect" their children, and their control diminishes as they become older and physically weaker and their offspring stronger and less dependent on parental investment (Apostolou, 2007b). Lastly, control over mating produces evolutionary pressures on children to evolve adaptations, such as psychological manipulation of their parents to accomplish their goals concerning mate choice (Trivers, 1974).

Sexual Selection under Parental Choice in Human Societies

Sexual Selection under Parental Choice in Pre-Industrial Societies

In pre-industrial societies offspring depend on their parents and close kin for food, support and protection. Accordingly, we expect that parents in these societies dominate mate choice. Apostolou (2007b), in an attempt to examine whether this is the case, he coded the mating patterns of a sample of 190 foraging societies spread across the globe. In these societies, parents exercise considerable control over the mating behaviour of their offspring, as the most common type of marriage is arranged marriage, since in 70% of the societies in the sample marriages were arranged. Marriage as the result of free courtship is found only in a small number of cases (4.3%). In arranged marriage, men, usually fathers, regulate the selection process, but it is not rare for women to be also influential. Finally, there is still the possibility for offspring to exercise choice by escaping, divorcing or forming extramarital affairs.

Moreover, Apostolou (2010b) analysed data from the Standard Cross Cultural Sample and found that in agropastoral societies the most frequent marriage type is arranged marriage, and marriage arrangements are controlled by males. Additionally, in agropastoral societies women are more likely, compared to men, to have an arranged marriage and to be married at a younger age. Nevertheless, children can still exercise choice by employing various means, such as divorce and extramarital affairs. Moreover, comparisons between agropastoral and foraging societies revealed that parents exercise more control over the mating decisions of their children in the former than in the latter societies. Male parents are also more frequently reported to dominate marriage arrangements in agropastoral societies than in hunting and gathering ones.

These patterns of mating have significant evolutionary implications. More specifically, the separation between agropastoral and foraging pre-industrial societies resembles the two most important stages of human evolution: humans lived as hunters and gatherers for the most part of human evolution, than over the last 10,000 years, there was a shift to a mode of subsistence based on agriculture and animal husbandry (Lee & DeVore, 1968). Therefore, by studying modern pre-industrial societies, we can make valid inferences concerning mating patterns in ancestral human societies (Ember, 1978; Lee & DeVore, 1968). Thus, as parental choice is dominant in modern foraging societies, we can infer that it was also dominant in ancestral foraging ones (Apostolou, 2007b). This is supported by research based on phylogenetic analysis which endeavors to recreate the conditions of ancestral societies (Walker et al., 2011).

Moreover, it can be further inferred that the agricultural revolution at the onset of the Holocene (10,000 years ago) gave rise to a subsequent increase in the influence of parents, and especially male parents, over the mating choices of their children (Apostolou, 2010b). As a matter of fact, we do not need to make assumptions for pre-modern agropastoral societies, as data for the mating patterns in these societies is already available from historical sources. This evidence indicates that parents, primarily male ones, had been arranging the marriages of their offspring, and in particular of their daughters, in ancient Greece (Vrissimtzis, 1997), ancient Rome (Balsdon, 1962), Imperial China (Gernet, 1970) and in pre-Victorian England (Stone, 1990).

Sexual Selection under Parental Choice in Post-Industrial Societies

In contrast to pre-industrial societies, the extensive educational needs of technology-based societies require marriage to be delayed, in anticipation of the conclusion of one's training. For that reason, individuals are married later in life, at a period when they are relatively independent from their families. Moreover, social protection systems, such as the welfare state, police, human rights etc., alleviate individuals' dependence on their family. Similarly, the legal system prevents the use of physical force, so parents are restrained from freely imposing their will by means of physical punishment. Since parents cannot use their children's reliance on their investment and their physical strength as means of coercion, they cannot directly dominate their mating choices. Consequently, individuals in post-industrial societies have the benefit of autonomy and lack of restrictions concerning mate choice. However, parental behaviour has been formed by evolution in order to attempt to dominate the mating behaviour of children, which denotes that even if parents cannot do so directly, they will attempt to do so by employing indirect means. To begin with, parents regulate considerable resources, for which their children are not unconcerned thus, these can be used to affect mate choice. For instance, manipulation of inheritance rights can give parents, particularly rich ones, a lever for effective manipulation (Apostolou, in press). Moreover, parents also employ means, like 'cajolery, persuasion, appeals to loyalty, and threats' (Apostolou, in press; Sussman, 1953) to influence the mating behaviour of their offspring. In the same vein, contemporary Chinese parents cannot inflict their choices on their children, but they still persist to exercise the function of the facilitator through their own social systems (Ikels, 1985). Chinese parents in the USA attempt to create situation in which their children can get together with other Chinese children of a desirable background. For example, they may stage a barbecue when an eligible relative from out of state comes to visit (Ikels, 1985). Although the real potency of parental choice in post-industrial societies remains to be estimated, the evidence currently existing indicates that parents are influential over mating choices.

In-Law Preferences

Marriage adds a new member to the family unit, who contributes significantly to its survival and reproductive efforts. In-laws integrate into their new family, supporting its subsistence activities (hunting, gathering, animal herding etc.), offering political support through their family relations, physical support in case of external threats (raids, wars, feuds etc.), their genes, and so on. However, in-laws differ. For example, some are diligent while others are indolent; some enjoy vigorous health while others suffer from chronic illness; some come from wealthy and powerful families while others have humble origins. Consequently, since prospective in-laws vary in their qualities, they also vary in the advantages they can offer to parents.

This reveals that parents have been experiencing strong evolutionary pressures to evolve preferences that permit them to select those in-laws who are most valuable to them (Apostolou, 2007a, 2010a). In contrast, parents who do not share such preferences are just as likely to choose a hard-working in-law as a lazy one. Therefore, the former gain a selective advantage over the latter, successfully augmenting the frequency of the genes that confer the disposition for these preferences in the population (Apostolou, 2007a). The result of this process is that parents have been equipped with well-defined in-law preferences. Given that not all the traits in an in-law provide parents with the same benefits, parents are likely to have stronger preferences for qualities which are more beneficial for them. Moreover, parental preferences are expected to be dependent upon the sex of the in-law. Specifically, the division of labour assigns or attributes diverse tasks and roles to men and women in a given society. For instance, in all recorded human cultures, men dominate over women in controlling resources (Whyte, 1978). In modern Western post-industrial societies, the wealthiest self-made individuals are men, while the wealthiest women are in the 'Rich List' because they inherited, or married into wealth. Accordingly, parents should value traits related to the capacity to acquire resources more in a son-in-law than in a daughter-in-law.

Two studies have been conducted to date which endeavour specifically to identify parental preferences. To begin with, Apostolou (2007a) requested from a group of British parents to rate the desirability of several traits in a prospective son-in-law and daughter-in-law. Parents appeared to have a hierarchy of preferences in which specific traits were deemed to be more essential than others. In particular, traits such as kind and understanding, good health, and good earning capacity were located at the top of the parental hierarchy, followed by traits such as education/intelligence, good cook/housekeeper, and good family background located in the middle of the parental hierarchy. Traits such as physically attractive and chastity were located at the bottom of the parental preference hierarchy. Moreover, parents rated traits in a different way in a daughter-in-law and a son-in-law. In particular, qualities such as ambition/industriousness, good financial prospects, and wealth were preferred more in a son-in-law than in a daughter-in-law, while traits, such as good housekeeper and good looks were preferred more in a daughter-in-law than in a son-in-law. Finally, both mothers and fathers were found to have the same opinion in terms of what they seek in an in-law of either sex. The second study, using anthropological evidence from 67 pre-industrial societies, identified 13 desirable qualities that parents seek in an in-law (Apostolou, 2010a). Among the most frequently reported traits were good character, good family background, industry and good worker, and subsequently favourable social status, wealth, and similar family social status. The least commonly reported traits were good looks and chastity. It was also found that traits such as good economic prospects, wealth and favourable social status were regarded highly more in a son-in-law than a daughter-in-law, while chastity was regarded highly more in a daughter-in-law than a son-in-law. Although not purposely intended to pinpoint in-law preferences, other studies provide information on the traits parents desire in an in-law. More specifically, Hynie et al. (2006) investigated in-law preferences of Chinese immigrants in North America and results reveal that good social status and understanding are among the most preferred traits. Borgerhoff Mulder (1988) reported that, among the pastoral Kipsigies in Kenya, parents prefer as sons-in-law individuals who enjoy high social status, are wealthy, educated, have a good character and are diligent. Yu, Proulx, and Shepard (2007) found that Matsigenka women in Peru prefer men with masculine faces as sons-in-law and interpreted this finding as a preference for good providers, as masculine men are, on average, perceived as better resource providers. Finally, Apostolou (2007b) found that among foragers, parents seek sons-in-law who are good hunters and have a good family background, while they prefer daughters-in-law who are industrious and come from good families. The knowledge of parental preferences, and how these deviate from the preferences of their children, can provide us with an understanding of the prevalence of certain mating strategies in the population.

MALE MATING STRATEGIES

Polymorphism in Mate Choice

Men seek to gain reproductive access to women, but the gatekeepers to this access are other men (usually fathers) who search for sons-in-law with particular traits. This creates substantial evolutionary pressures for men to attract interest as in-laws from parents and particularly from male parents. Nevertheless, women's mate choices are not entirely restricted by their parents, since they can engage in informal relationships before marriage and in extramarital relationships subsequent to marriage, while they can also divorce the spouses their parents have chosen for them (Apostolou, 2010b). Consequently, there is freedom for women to exercise choice, which delineates that there have also been evolutionary pressures on men to discover wide-ranging ways to appeal to women as mates.

To put it differently, there have been at least two mating niches that men can address: The parental niche (being selected through parental choice) and the female niche (being selected through female choice). The evolutionary effect of the existence of two niches is that men may have evolved to attend to both parental choice and female choice, or some men may have evolved to attend to parental choice and others to attend to female choice.

The latter scenario is more likely because, in cases where a species occupies multiple niches, a polymorphic equilibrium (e.g., two or more distinct specializations) is usually more optimal than a monomorphic one (e.g., a single specialization), as specialists are more efficient than generalists (Wilson, 1994). This is due to the fact that a male specializing in appealing to parental choice obtains reproductive advantages from marriage without having to bear the cost of appealing to female choice. A male specializing in appealing to female choice may obtain reproductive advantages by engaging in different casual relationships without bearing the costs of appealing to parental choice.

As a species we are a polymorphic: Some individuals have blond hair, while others have dark hair; some individuals have blue eyes and others have green eyes. We are also polymorphic in terms of behavior. Some individuals are hard-working and others are lazy; some are kind and others are evil and so on. Approximately half of the variations between people with respect to personality are due to the fact that they share diverse genes (Plomin et al., 2008). Accordingly, polymorphism, being frequent in our species, should also exist in mating behavior, mainly in light of the fact that there is more than one mating niche.

Our knowledge of in-law and female preferences permits us to hypothesize how the two male morphs are likely to be structured. To begin with, men specializing in attracting interest from parents are likely to give emphasis to their personality and family background, and pay less attention to their physical appearance. They are also likely to be concerned about impressing and gaining the respect of other men. Conversely, men specializing in attracting the interest of women are likely to accentuate more their physical appearance, being charming, but pay less attention in communicating their good family background.

Assuming that parental choice has been dominant throughout human evolution, we would expect that the male morph that addresses parents should have been much more frequent than the male morph that addresses women. To put it in a different way, the parental niche has been much bigger than the female niche, leading to an equilibrium where the majority of men specialize in attracting interest from parents, and a minority in attracting interest from women.

Overall, men are likely to have evolved two strategies in order to obtain access to the opposite sex, one through females themselves and one through their parents. It is also of paramount importance to stress that there is a third approach, which permits men to force mating access to women.

THE EVOLUTION OF RAPE

Circumventing Parental Choice

When they look for sons-in-law, parents are in search for men who are endowed with qualities which are advantageous for them. Evidence from modern and historical societies reveals that parents are interested in finding sons-in-law who have a high resource-generating capacity, have a good family background, control wealth, and have a good character (Apostolou, 2010a, 2012; Borgerhoff Mulder, 1988; Koster, 2011). Men who lack desirable qualities, such as social status and find themselves in a context where mating is regulated, endure a substantial reproductive loss, since they are inclined to be excluded from mating or have to compromise by settling for a woman of low mate value.

To be more precise, parents with daughters of elevated mate value (e.g., attractive, young, etc.) are not going to be willing to give them as wives to men of an inferior mate value (e.g., men who lack desirable traits). Therefore, if a man of inferior mate value seeks to gain parental approval, he needs to address parents whose daughters also have an inferior mate value (e.g., are older, unattractive, have children from previous marriages, etc.), and who are thus less reluctant to accept him as a son-in-law. In fact, this man will endure substantial fitness losses, as he will have to either settle for a woman of low mate value or opt out from the reproductive process.

In addition to parental choice, there are two related factors which also operate to alleviate the ability of men, who have a low mate value, to attract desirable wives. One is polygyny, which is practiced in most of the pre-industrial societies and acts to exclude low status men from gaining reproductive access (Blood, 1972). The reason is that high mate value men are able to attract multiple wives, leaving those at the bottom of the hierarchy single. In the same vein, hypergyny, that is, women marrying up the social hierarchy, a common practice in many societies, pulls women out of the lower classes, leaving a lot of low-status men without wives (Boone, 1986).

In this context, then, a forced-sex mating strategy can decrease the reproductive costs that a man would probably suffer by circumventing parental choice, permitting sexual access to women of high mate value, women that these low mate value men could not have accessed in any other way. This predicts that rapists are mostly young men of low mate value. This prediction is consistent to the findings of several studies (Thornhill & Palmer, 2000; Thornhill & Thornhill, 1983). This predicts also that the victims of rape should predominantly be high mate value women, who would not be accessible by these rapists in any other way. Consistent with this prediction, the majority of rape victims are frequently young women at the peak of their fertility (Greenfield, 1997; Kilpatrick et al., 1992; Thornhill & Palmer, 2000).

Circumventing Female Choice

In societies where mating is controlled, women still have some space to exercise mate choice. For instance, they can exercise mate choice by forming extramarital relationships or taking a divorce and getting married later, when their parents are no longer alive (Apostolou, 2010b). Consequently, a man can address female choice directly. However, women are in search of men with desirable traits, such as high resource acquisition capacity and physical attractiveness (Buss, 2003). This indicates that a man who lacks these traits is unlikely to be able to successfully appeal to female choice if he addresses high mate value women. He will be more effective, however, if he addresses low mate value women, such as older ones who are less likely to receive offers from men of high mate value. Nevertheless, getting married to an older woman will not enhance his fitness much, as the reproductive years of that woman have already been or will soon be exhausted.

A man will inevitably need to address female choice within marriage. In particular, a man's mate value may decrease if for instance he suffers an injury from a fight, or experiences a status loss. This will make his wife perceive him as a less desirable mate. Additionally, in a context where mating choice is regulated, a man may effectively pass through parental choice, but this does not mean that he will instantly obtain sexual access to his wife, as the latter may decline her husband's requests to permit him to gain sexual access.

Consequently, a forced-sex mating strategy can offer fitness benefits by permitting men to circumvent female choice. The female choice within marriage cases predicts further that rape would also occur within marriage. Consistently with this prediction, between 10% and 26% of women report experiencing marital rape (Russell, 1990; Watts, Keogh, Ndlovu, & Kwaramba, 1998). It is further predicted that rape will be more likely to occur when the mate value of the husband diminishes (e.g., he loses his job) or the mate value of the wife augments (e.g., she loses weight).

Finally, in societies which practice arranged marriage, rape within marriage enhances the strength of parental choice as a sexual selection force because it makes the choices of parents on their daughters consequential. This predicts that in these societies, the parents of a man may prompt him to force sex on his wife if she rejects his requests to obtain sexual access, and the parents of the bride may also encourage or at least accept such action. This prediction also remains to be addressed by future research.

In a pre-industrial context, a forced-sex mating strategy can endow substantial reproductive advantages to inferior mate value men by enabling them to circumvent parental and female choice. This strategy is not of course cost-free, otherwise nearly all men would adopt it most of the times. In particular, a rapist is likely to meet the fierce reaction of a woman's parents and other kin, he is likely to meet the resistance of the woman, which might result in physical injury, and he is likely to face the vengeance of the woman, particularly in the case of marital rape.

CONCLUSION

The model of parental choice predicts that for the greatest part of the period of human evolution, parental choice constituted a significant sexual selection force, with individual mate choice still likely to be exercised. This prediction is consistent with evidence from the anthropological and historical records which demonstrate that across human pre-industrial societies mate choice is regulated, but there is also room for individuals to exercise their own choice.

Based on the abovementioned, it has been argued that men have evolved at least three diverse strategies in order to obtain sexual access to the opposite sex: they can address female choice, parental choice, or attempt to circumvent both selection mechanisms by force i.e., through rape. Future research needs to explore each strategy and assess their relevant fitness payoffs.

REFERENCES

Apostolou, M. (2007a). Elements of parental choice: the evolution of parental preferences in relation to in-law selection. *Evolutionary Psychology*, *5*, 70-83.

Apostolou, M. (2007b). Sexual selection under parental choice: the role of parents in the evolution of human mating. *Evolution and Human Behavior*, *28*, 403-409.

Apostolou, M. (2008). Parent-offspring conflict over mating: The case of beauty. *Evolutionary Psychology*, *6*, 303-315.

Apostolou, M. (2010a). Parental choice: What parents want in a son-in-law and a daughter-in-law across 67 pre-industrial societies. *British Journal of Psychology*, *101*, 695-704.

Apostolou, M. (2010b). Sexual selection under parental choice in agropastoral societies. *Evolution and Human Behavior*, *31*, 39-47.

Apostolou, M. (2011). Parent-offspring conflict over mating: Testing the tradeoffs hypothesis. *Evolutionary Psychology*, *9*, 470-495.

Apostolou, M. (2012). Sexual selection under parental choice: Evidence from sixteen historical societies. *Evolutionary Psychology*, *10*, 504-518.

Apostolou, M. (in press). Do as we wish: Parental tactics of mate choice manipulation. *Evolutionary Psychology*.

Balsdon, J. P. V. D. (1962). *Roman women: their history and habits*. London: The History Book Club.

Blood, R. O. (1972). *The family*. New York: The Free Press.

Boone, J. L. III (1986). Parental investment and elite family structure in preindustrial states: A case study of late medieval-early modern Portuguese genealogies. *American Anthropologist*, *New Series*, *88*, 859-878.

Borgerhoff Mulder, M. (1988). The relevance of the polygyny threshold model to humans. In C. Mascie-Taylor & A. Boyce (Eds.), *Human mating patterns* (pp. 209-230). Cambridge: Cambridge University Press.

Buss, D. M. (2003). *The evolution of desire: Strategies of human mating* (2nd ed.). New York: Basic Books.

Buunk, A. P., Park, J. H. & Dubbs, S. L. (2008). Parent-offspring conflict in mate preferences. *Review of General Psychology*, *12*, 47-62.

Darwin, C. (1871). *The descent of man and selection in relation to sex.* London: Murray.

Ember, C. R. (1978). Myths about hunter-gatherers. *Ethnology*, *17*, 439-448.

Gernet, J. (1970). *Daily life in China on the eve of the Mongol invasion 1250-1276*. Stanford: Stanford University Press.

Greenfield, L. (1997). *Sex offenses and offenders: An analysis of data on rape and sexual assault*. Washington, DC: Bureau of Justice Statistics, U.S. Department of Justice.

Hynie, M., Lalonde, R. N. & Lee, N. (2006). Parent–child value transmission among Chinese immigrants to North America: The case of traditional mate preferences. *Cultural Diversity and Ethnic Minority Psychology*, *12*, 230–244.

Ikels, C. (1985). Parental perspectives on the significance of marriage. *Journal of Marriage and the Family*, *47*, 253-264.

Kilpatrick, D., Edmunds, C. & Seymour, A. (1992). *Rape in America: A report to the nation*. Arlington, VA: National Victim Center.

Koster, J. (2011). Hypothetical rankings of prospective husbands for female kin in lowland Nicaragua: Consensus analysis indicates high agreement and associations with wealth and hunting skill. *Evolution and Human Behavior*, *32*, 356-363.

Lee, R. B. & Devore, I. (1968). *Man the hunter*. New York. Aldine.

Miller, G. (2000). *The mating mind*. London: BCA.

Perilloux, C., Fleischman, D. S. & Buss, D. M. (2008). The daughter guarding hypothesis: Parental influence on, and emotional reaction to, offspring's mating behavior. *Evolutionary Psychology*, *6*, 217–233.

Perilloux, C., Fleischman, D. S. & Buss, D. M. (2011). Meet the parents: Parent-offspring convergence and divergence in mate preferences. *Personality and Individual Differences*, *50*, 253-258.

Plomin R., DeFries, J. C., McClearn, J. C., McGuffin, P. (2008). *Behavioral genetics* (5th ed.). London: Worth Publishers.

Russell, D. E. H. (1990). *Rape in marriage* (Rev.ed.). Indianapolis: Indiana University Press.

Stone, L. (1990). *The family, sex and marriage in England* 1500-1800. London: Penguine Books.

Sussman, M. B. (1953). Parental participation in mate selection and its effect upon family continuity. *Social Forces*, *1*, 76-81.

Thornhill, R. & Gangestad, S. (1993). Human facial beauty: Averageness, symmetry and parasite resistance. *Human Nature*, *4*, 237–269.

Thornhill, R. & Palmer, C. T. (2000). *A natural history of rape: Biological bases of sexual coercion*. Cambridge, Massachusetts: The MIT Press.

Thornhill, R. & Thornhill, N. (1983). Human rape: An evolutionary analysis. *Ethology and Sociobiology*, *4*, 137-173.

Trivers, R. L. (1972). Parental investment and sexual selection. In B. Campell (Ed.), *Sexual selection and the descent of man: 1871-1971* (136-179). Chicago: Aldine.

Trivers, R. (1974). Parent offspring conflict. *American Zoologist*, *24*, 249-264.

Vrissimtzis, N. A. (1997). *Love, sex & marriage in ancient Greece* (3rd ed.). Athens: Polygrama.

Walker, R. S., Hill, K. R., Flinn, M. V. & Ellsworth, R. M. (2011). Evolutionary history of hunter-gatherer marriage practices. *PLoS ONE, 6*, e19066.

Watts, C., Keogh, E., Ndlovu, M. & Kwaramba, R. (1998). Withholding of sex and forced sex: Dimensions of violence against Zimbabwean women. *Reproductive Health Matters, 6*, 57–65.

Whyte, M. K. (1978). *The status of women in preindustrial societies*. Princeton, NJ: Princeton University Press.

Wilson, D. S. (1994). Adaptive genetic variation and human evolutionary psychology. *Ethology and Sociobiology*, *15*, 219-235.

Yu, D., Proulx, S. R. & Shepard, G. H. (2007). Masculinity, culture and the Paradox of the Lek. In V. Swami & A. Furnham (Eds.), *Body beautiful: Evolutionary and sociocultural perspectives* (88-107). London: Palgrave Macmillan.

Zahavi, A. (1975). Mate selection: A selection for a handicap. *Journal of Theoretical Biology*, *53*, 205-214.

Zahavi, A. & Zahavi, A. (1997). *The handicap principle: A missing piece of Darwin's puzzle*. New York: Oxford University Press.

INDEX

D

E

I

K

L

M

Q

R

S

T

U

V

W

X

Y